# THE LITTLE BROWN BOOK

Published by the Case Alumni Association
in support of the
Case School of Engineering
of
Case Western Reserve University

Dedicated to
*Roger H. Cerne*
in honor of his 25 years of service to the
Case Alumni Association

Based on the book
*Wonders in Numbers*
by
Anthony J. Schneider

THE LITTLE BROWN BOOK

Copyright © 1999 by Anthony J. Schneider, San Clemente, CA. All Rights Reserved. Printed in the United States of America. No part of this publication may be reproduced, stored in a retrieval system, or transmitted in any form or by any means, electronic, mechanical, photocopying, recording or otherwise, without the prior written permission of the publisher.

ISBN 0-9672834-0-X

Book design by Diane M. Smith

First edition – June 1999
Based on *Wonders in Numbers* © 1994

# CONTENTS

Preface — vi

**Chapter 1**    A Space Odyssey — 1
*Earth's speeds in Universe – Distances in space – Time as a measure of distance – Viewing history in the making*

**Chapter 2**    How We Count — 6
*Decoding Roman and Arabic numerals – Number bases made easy*

**Chapter 3**    Taking a Chance — 11
*Probability – Chance – Odds*

**Chapter 4**    From Here to Infinity — 18
*Number sets – Finite numbers – Concept of infinity*

**Chapter 5**    The Right Size — 21
*Importance to animal life – Optimum size – Relevance to space travel*

**Chapter 6**    Planets in Motion — 26
*Sun vs. Earth as center of revolution – Planetary paths – Gravity*

**Chapter 7**    Let's Talk Percentages — 31
*Percent vs. incremental value as measure of change – Thinking in terms of percentages – Logarithmic scale*

**Chapter 8**    A Number is a Number, or is It? — 37
*Basic types of numbers – Evolution of multiplication and division – Complex numbers – Special sets of numbers*

**Chapter 9**    Scientific Notation — 43
*Simplified method for handling large numbers – Ratio prefixes – Greek alphabet*

Chapter 10     Mean, Median, Mode     45
*Arithmetic mean – Median – Mode – Percentile – Weighted and moving averages – Geometric mean*

Chapter 11     Pi All Over     51
*Estimating the value of Pi – Radian – Polar coordinates*

Chapter 12     Destination Moon     55
*Launch – Escape velocity – Parking orbit – Lunar trajectory – Moon landing*

Chapter 13     Radiation All Around     62
*Electromagnetic spectrum – Radiation waves – Forms of radiation*

Chapter 14     Where's the Ether!     70
*Ether Theory – Speed of light – Michelson-Morley experiment*

Chapter 15     Mendeleev's Numbers     76
*Discovery and classification of chemical elements – Atomic number – Atomic weight – Particles of matter – Chemistry*

Chapter 16     *e*-Gads     87
*Linear vs. geometric series – Interest compounding – Exponential growth – Carbon dating – Natural logarithms*

Chapter 17     Mathematics in Art     93
*Measure of aesthetics – Perspective drawing – Golden section*

Chapter 18     Go Gaussian     97
*Theory and application of the bell-shaped curve*

Chapter 19     Let's Take a Poll     103
*Why polls are conducted – Margins of error – Making polls honest*

Chapter 20     Computer Numbers     108
*Hexadecimal numbers – Number resolution*

Chapter 21     More on Averages     113
*Averaging techniques for dynamic data – RMS value*

## Contents

**Chapter 22  Zany World of Weights and Measures  117**
*Units of length and weight – Temperature scales – Absolute temperature – Metrification*

**Chapter 23  Secrets of a Cone  125**
*Cone – Circle – Ellipse – Parabola – Hyperbola – Occurrences in nature*

**Chapter 24  Longitude, Oh, Longitude  135**
*Greenwich Mean Time – Nautical mile – Chronometer*

**Chapter 25  Equations – Our Best Friends  141**
*Equations – Graphs – Tables – Empirical Data – Accuracy*

**Chapter 26  Enigmas of Space, Time and Matter  148**
*Relativity in our daily lives – Einstein's relativity – Mass/energy – Fission and fusion – Cold fusion*

**Chapter 27  Is Anyone Out There?  157**
*Prospects for life on other worlds – UFO challenges – Corona Incident*

**Chapter 28  From Chaos to Order  164**
*Order in our Universe – Chaos – Nonlinearities*

**Chapter 29  Demon dB  169**
*Bel/decibel code – Voltage vs. power – Richter scale*

**Chapter 30  Frequency Analysis  175**
*Music synthesis and analysis – Fourier analysis*

**Chapter 31  Calculus 001  179**
*Introduction to integration and differentiation*

Conversion Tables  185

Suggestions for Further Reading  191

Index  193

# PREFACE

Several years ago my friend and colleague, Tony Schneider, gave me a copy of his book *Wonders in Numbers* which he published in 1994. In reading this delightful little book I was struck by the hundreds of "gee, I didn't know that" relationships between numbers and the physical/political world in which we live. Tony originally intended his book to stimulate student interest in math and science. I felt that the potential audience could be greatly expanded. Since the Case Alumni Association is always looking for mechanisms in which to increase its visibility, I suggested that a special edition of *Wonders in Numbers* might make a fine promotion piece.

*The Little Brown Book* is the culmination of this effort. It is hoped that it will attract interest in the Case School of Engineering of Case Western Reserve University. Every student at Case will find useful information in this book that will be of benefit to their undergraduate education. For everyone else, you don't have to be a rocket scientist to glean some wonderful gems of knowledge from this book that you can use to amaze your friends and associates.

**The Case School of Engineering.** Case Western Reserve University was created in 1967 by the merger of Case Institute of Technology (founded in 1880 as Case School of Applied Science) and Western Reserve University (founded in 1826). The federation of the two schools created a comprehensive university, rich in liberal arts and technological disciplines. The Case School of Engineering offers one of the premier engineering programs in the country. It is known for rigorous and challenging curricula, extensive research activities, outstanding faculty and for the achievements of its graduates.

**The Case Alumni Association.** Founded in 1885 by the first five graduates of the Case School of Applied Science, the CAA is now in its 114th year of continuous operation. It is the oldest independent alumni association of engineering and applied science graduates in the country. Membership in the CAA is conferred upon all Case graduates. Today, the Association serves the interests of over 16,000 Case alumni.

The purpose of the Case Alumni Association is to promote the traditions of Case; the priorities of the Case School of Engineering in particular, and Case Western Reserve University in general. The

Case Alumni Association raises monies from alumni and other friends through its CaseFund® Annual Giving Program, reaching all Case alumni from 1909 to 1999. This money is then used for the benefit of Case in such areas as undergraduate laboratory equipment, scholarships, fellowships, libraries, student prizes and awards, support to student groups and other priorities of Case.

**Case Alumni Association Information Services.** In addition to this book the CAA publishes the *Case Alumnus*, a quarterly magazine that is sent to all members of the CAA, Case students, and other interested individuals. The CAA offers Case Clubs in different cities throughout the United States. These clubs allow alumni to meet socially to establish friendships and business connections, while staying informed about the latest campus developments. Officers and staff from the CAA visit each club to provide interesting programs and dinners for Case Club guests.

The Case Alumni Association also offers the Annual All-Classes Reunion for alumni celebrating a five year anniversary. Alumni are encouraged to return to Cleveland to participate in special class events as well as opportunities to visit the campus and socialize with alumni of all years.

It is fitting that *The Little Brown Book* has been published on the occasion of the 45th reunion of the Case Institute of Technology class of '54. On behalf of the Case Alumni Association and all of the individuals who assisted with this publication, we hope that you will enjoy the book and find it useful in your personal activities over the years.

Jack K. Mowry, CIT '54

**Case Alumni Association**
107 Crawford Hall
10900 Euclid Avenue
Cleveland, OH 44106-7073
216-231-4567
FAX 216-368-4714
e-mail casealum@po.cwru.edu
http://www.caa.cwru.edu

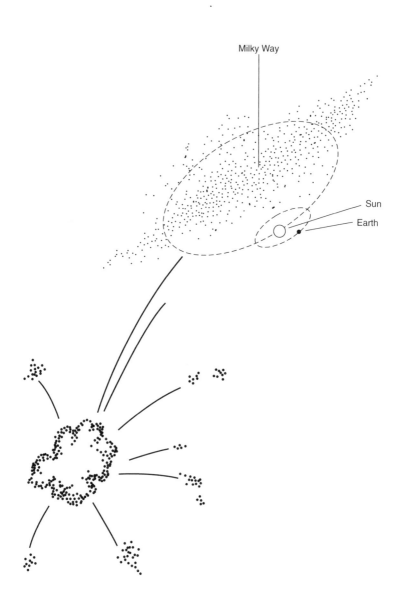

# 1
# A SPACE ODYSSEY

Imagine yourself on a giant spacecraft. For you are, you know. Our planet Earth is a gigantic spaceship whirling endlessly it seems through the vast emptiness of the universe. And we are its passengers.

The numbers involved in describing our travels are absolutely mindboggling. First of all, Earth is rotating on its axis. On the equator, where speed is greatest, our speed is:

Rotation of Earth – 1000 MPH (at equator)

You can calculate this value yourself. The circumference at the equator is approximately 24,000 miles, and Earth turns through this distance every 24 hours for a speed of about 1000 miles per hour. Most of our spaceship's population, however, lives in the middle latitudes where the circumference is smaller and the speed therefore slower. We'll say our speed is about 800 MPH.

Secondly, Earth is orbiting the sun at the rate of one revolution per year with an average speed of:

Orbiting the sun – 65,000 MPH

We passengers depend upon the sun for light and heat. Without the sun we would all perish.

Thirdly, the sun doesn't just hang motionless in the middle of nowhere. It rotates on its own axis while orbiting along with more than 100 billion other stars around the center of our galaxy, which we call the Milky Way. This gives us an additional speed:

Following the sun's orbit – 650,000 MPH

It takes the sun approximately 240 million years to orbit the Milky Way. During its estimated 4.6 billion years of existence, the sun has completed about 20 orbits. Astronomers predict it is about half way through its life, after which it will slowly swallow our spaceship in a huge ball of fire. How many miles will the sun have traveled by then? An astronomical number, isn't it? But, in numbers of orbits, it is an earthly number like 40. What an incredible fact!

Finally, our galaxy is also moving. If the Big Bang theory is correct, the Milky Way is hurtling outward from its source of life many eons ago. This speed must also be very high. Herein lies the frontier of our knowledge:

Following the Milky Way – Speed Unknown

So the next time you gaze at the nighttime sky, you can begin to understand the enormity of the universe if you contemplate the speeds you are traveling in space: 800 MPH Earth rotation, 65,000 MPH orbiting the sun, 650,000 MPH following the sun around our galaxy, plus an unknown speed following our galaxy.

## THE VASTNESS OF THE UNIVERSE

With trillions of stars and galaxies in the universe, the distances are so vast that it is impractical to measure them in kilometers or miles. We need a whole new concept of thinking.

Astronomers have devised the "light-year" as their unit for measuring distance. It is based on the speed of light which is theoretically the maximum speed at which energy can travel. The speed of light is 186,282 miles per second. Light coming from the sun and stars travels at this speed. A light-year is the distance light travels in one year. Multiply it out (assuming 365.25 days/yr):

186,282 mi/sec × ___sec/hr × ___hr/da × ___da/yr = 5,878,612,843,000 mi/yr

**You can see that:**

1 light-year ≈ 6 trillion miles

Yet, even a light-year is a very small unit of measurement for distant galaxies. Consider the distances in this table.

| Distance from Earth (Light-Years) | |
|---|---|
| Moon | 0.00005 |
| Sun | 0.00155 |
| Alpha Centauri (Sun's Nearest Star) | 4.2 |
| Sirius (Our Brightest Star) | 8.8 |
| Polaris (Pole Star) | 820 |
| Center of Milky Way | 28,000 |
| Far Edge of Milky Way | 78,000 |
| Andromeda Galaxy | 2,400,000 |
| Most Distant Galaxy Known | 10,000,000,000 |

Are you beginning to appreciate not only the immensity of the universe, but how relatively close we are to the sun? It only takes about 8 minutes for light to travel from sun to Earth.

Note that the most distant galaxy known is about 10 billion light-years away. That's about 60 billion trillion miles. How far does the universe extend beyond that? And how far in other directions? Are there really any limits? These are among the many things astrophysicists are still studying.

They say that the breadth of the universe observable with today's technology is about 160 billion light-years, or $10^{24}$ miles. This distance is impossible to comprehend. Consider the following. The Big Bang occurred some 15 billion years ago. An immense length of time. Since then, approximately $10^{17}$ seconds have elapsed. Yet you have to multiply this number by $10^7$ (i.e., 10 million) to obtain the number of miles that just the known universe spans. Wow!

## VIEWING HISTORY

A light-year is the distance light travels in one year, correct? If a star is one light-year away, its light takes one year to reach Earth. Note that we can use time to measure distance. Is this a new concept? Probably not. After all, if someone asked you how far it is to your grandmother's home, you might say "a half-hour drive."

Now consider this. When we look at Polaris, which is 820 light-years away, we see light that Polaris emitted 820 years ago. In other words, we are seeing Polaris as it existed in the $12^{th}$ century. We are seeing history in the making, are we not? If Polaris had experienced a violent event at that time in history, we would just be seeing it today. Moreover, we don't know if Polaris still exists. We won't know until 820 years from now when light emitted today reaches Earth! Our observations of Polaris will forever have this time lag. Amazing, isn't it?

Astronomers analyze visible light and radio-frequency emissions to investigate the makeup of stars and to learn more about how stars originate and how they die. They expand their capabilities by building ever more powerful telescopes and by locating them in ever more remote areas to eliminate interference from background light. The darker the environment, the more stars they can see. We recently made significant advancements by orbiting telescopes in space, where there is no background light. This also enables astronomers to analyze infrared, ultraviolet and microwave energies not seen before because they are absorbed by Earth's atmosphere.

The farther we can see, the earlier is the history that we can view. In 1989, astrophysicists in Arecibo, Puerto Rico, discovered a galaxy forming 65 million light-years from Earth. They are seeing history as it happened 65 million years ago. Our hope for the Hubble Space Telescope is to see back 13 to 15 billion years to the edge of time when the entire universe was young. By comparing these events to more recent events in stars closer to us, we will be better able to understand how the universe began and how to predict its future.

## 2
# HOW WE COUNT

Soon after we learned to say "Mommy" and "Daddy," our parents taught us to count from one to ten. From there we went on to larger and larger numbers. Here is an opportunity to better understand our counting system, how efficient it is and how difficult arithmetic might have been for students of another era.

Counting techniques date back to early Chinese and Egyptian civilizations some 5000 years ago. They had much in common with later "Roman" numerals, which are still in limited use today. We see them on clock faces, cornerstones of buildings, prologues to books and movies and the numbering of Super Bowl games.

All counting systems employ symbols to designate integer values. The symbols are actually a code. The Romans used letters of the alphabet for their code:

### ROMAN NUMERALS

| M | D | C | L | X | V | I |
|---|---|---|---|---|---|---|
| 1000 | 500 | 100 | 50 | 10 | 5 | 1 |

Let's decode this number:

M DC LX VI

$$5 + 1 = 6$$
$$50 + 10 = 60$$
$$500 + 100 = 600$$
$$1000 = 1000$$
$$\overline{1666}$$

Since the Romans had symbols for only 1 & 5, 10 & 50, 100

& 500, etc., they had to repeat them to obtain intermediate values. For example, their code for the one's digit was:

| | | |
|---|---|---|
| I = 1 | IIII = 4 | VII = 7 |
| II = 2 | V = 5 | VIII = 8 |
| III = 3 | VI = 6 | VIIII = 9 |

We refine their code a bit and write IV for 4 and IX for 9, putting the I in front of the next higher value to indicate subtraction. This is shorter than writing IIII and VIIII. Nevertheless, the string of symbols can be quite long. And it gets worse when we add new symbols for larger numbers. But in those days there may have been little need for large values.

Try some practice problems:

M DCC LXXX IV = 1784
MMM CD LX VII = ____
MMDCCXLVI = ____

Imagine how tedious it would be if you had to use Roman numerals. How would you multiply and divide? Fortunately, our system is much more efficient. We use Arabic numerals in a system developed in India about 1200 years ago, translated later by the Arabs and brought to Europe about 1000 years ago.

ARABIC NUMERALS

0, 1, 2, 3, 4, 5, 6, 7, 8, 9

We code our numbers as follows:

```
1 9 6 4 2
│ │ │ │ └── 2  ×  1     =      2
│ │ │ └──── 4  ×  10    =     40
│ │ └────── 6  ×  100   =    600
│ └──────── 9  ×  1000  =   9000
└────────── 1  ×  10000 =  10000
                          ─────
                          19642
```

Our code is shorter and easier to interpret than the Roman

code. But precisely what is it that makes our system superior? First of all is the fact that our number of symbols, which is 10, matches our number base of ten. This gives us a unique symbol for each integer, 0 through 9. We do not have to repeat or mix symbols: 2 is 2, not II, 8 is 8, not VIII, etc. The most important difference is that we have the symbol '0,' and the Romans did not. We can write:

$$
\begin{array}{r}
9\ 0\ 5 \\
5 \times 1 = 5 \\
0 \times 10 = 0 \\
9 \times 100 = \underline{900} \\
905
\end{array}
$$

The '0' tells us there are "no tens" in this number. Do you see that the Romans had no direct way to say "no tens?" This is why they used separate sets of symbols for 1s, 10s, 100s, etc. To indicate there were no tens, they omitted L and X. Since we have a zero, we don't need extra symbols. We reuse our original symbols, 0 through 9, for each position. This makes our numbers more concise and easier to decode.

## OTHER NUMBER BASES

Something else very interesting is going on here. Let us recall a number previously decoded. We can write it as an equation:

$19642 = 1 \times 10000 + 9 \times 1000 + 6 \times 100 + 4 \times 10 + 2 \times 1$

or, better yet:

$19642 = 1 \times 10^4 + 9 \times 10^3 + 6 \times 10^2 + 4 \times 10^1 + 2 \times 10^0$

The equation tells us how many times each power of ten appears in the number. Observe that we prefer to use $10^0$ in place of '1' because it makes the equation uniform. Any number raised to the zero power equals 1. We can demonstrate this algebraically:

$$a = \frac{a^1}{a^1} = a^{1-1} = a^0$$

Here is what is so interesting. As long as the number of symbols (always starting with zero) is identical to the number base, we have a counting system that overcomes the weaknesses of the Roman system. We do exactly this in computers, where we need a code that uses only two symbols. This is because computers process only YES/NO information. Therefore, we select a base-of-2 and use the symbols 0 and 1. We call it the "binary" code. It is not new, having been referred to in a Chinese book written about 3000 B.C. Here is an example of a binary number and how it is coded:

100111101

| | | | | |
|---|---|---|---|---|
| 1 | × | $2^0$ | = | 1 |
| 0 | × | $2^1$ | = | 0 |
| 1 | × | $2^2$ | = | 4 |
| 1 | × | $2^3$ | = | 8 |
| 1 | × | $2^4$ | = | 16 |
| 1 | × | $2^5$ | = | 32 |
| 0 | × | $2^6$ | = | 0 |
| 0 | × | $2^7$ | = | 0 |
| 1 | × | $2^8$ | = | 256 |
| | | | | 317 |

The code simply tells how many times (0 or 1) each power of two appears in a number. These powers of two are called binary numbers. The great Scottish mathematician, John Napier, spoke with interest of the fact that a farmer could weigh grain by loading a scale's balance pan from a set of weights of 1, 2, 4, 8, 16, . . . pounds.

You can just as easily construct a code for the base-of-3. You know that you must use three symbols (0, 1, 2) to tell how many times each power of three is contained in any given number. Above, we wrote 317 in the base-of-2. Here it is in

the base-of-3:
102202

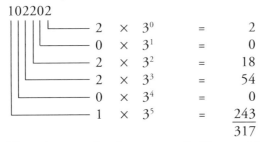

Try other number bases as well. You could even choose a base greater than ten, couldn't you? But you would have to design some new symbols, wouldn't you? If you chose a base-of-12, what symbols would you use for the two numbers that follow 9? They must be simple and must not be confused with any other mathematical symbol.

We marvel at the sophistication shown by ancient civilizations. We often hear someone say that we count from one to ten because our ancestors counted on their ten fingers. That might be true, but it took a lot of insight to achieve an efficient system. After all, the Romans counted from one to ten, but their system was not efficient. It took people from far-off India to discover that we needed more symbols and that one of them had to be zero. This is said to be one of history's most important technical discoveries, ranking with fire and the wheel.

# 3
# TAKING A CHANCE

You are continually exposed to statistics dealing with highway safety, airline safety, health risks, even ads for lottery tickets. Statistics is not an easy subject, but with some basic knowledge of probability theory you can have fun evaluating events around you.

## PROBABILITY

The formal study of probability and statistics originated in the 17$^{th}$ century in response to questions about gambling. Our word "hazard" is derived from an Arabic word for "dice."

No matter if you are gambling with money or deciding whether it is safest to fly, drive or take a train, you are dealing with probabilities.

Let's start with something very basic. Ask your grandfather what gambling game he played when he was your age. He might say he tossed coins. One player tosses a coin and the other player calls heads or tails. A correct call wins a coin. An incorrect call loses one. This is a perfect place to begin our story.

If the coin is perfectly balanced, there is equal probability of its landing head $H$ up or tail $T$ up:

$$\text{Probability} = \frac{\text{Number of Favorable Outcomes}}{\text{Total Number of Outcomes}}$$

$$\text{Probability of a Head} = \frac{H}{H \text{ or } T} = \frac{1}{2} = 50\%$$

$$\text{Probability of a Tail} = \frac{T}{H \text{ or } T} = \frac{1}{2} = 50\%$$

What else do we know? Well, we know from experience that if we toss a coin repeatedly, it will not alternate *H, T, H, T, H, T*. Yet statistics tell us that heads and tails will come up equally often in the long run. This is the critical point – the long run. The more times we toss the coin, the closer we come to 50% heads and 50% tails. But what happens in between?

If you were calling heads and heads came up three times in succession, would you switch to tails? Probably not. What about five times in a row? Or ten times? Eventually your intuition begins to tell you that the next toss is more likely to be a tail than a head. But this cannot be true because we stipulated that a head or tail is equally likely on each toss!

This demonstrates a fundamental rule of gambling – long-run outcome does not affect individual events. You may still think there is some contradiction. But there is not. The explanation lies in the fact that "long run" could mean thousands of tosses. During this period of time, long runs of heads will be balanced out by long runs of tails. We can only guess what will happen in the short run. This uncertainty is what leads to gambling.

The question is whether gamblers have enough money to sustain a long run of losses before luck turns in their favor. Of course, they don't fall into this trap. Gamblers in Las Vegas and Atlantic City know they will lose if they play long enough at one time, if for no other reason than the casino has small odds in its favor due to its method of payout. Gamblers hope to hit short streaks of good luck and then get out when their luck begins to turn. They continually jump in and out of the game.

\* \* \*

Here is a mental block we all have. When we read that the outcome of a risky act will be unfavorable one time in a thousand, we tend to think of the event occurring toward the thousandth time. Not so, it is just as likely to happen the first time! Or several times in a thousand. Or not at all. Why? Because 1/1000 is merely the long-term probability. Think about this the next time you are tempted to risk your health or safety.

\* \* \*

During this writing, I watched a TV documentary on the famous B-17 Flying Fortresses that pioneered daylight bombing raids on Nazi targets during World War II. In the early months, the role of the flight crews was so dangerous that men were rotated home after 25 missions. How dangerous? Well, there was only a 1-in-3 chance of completing 25 missions. A survival rate of only 33%! This raises a very interesting question. Assuming all missions were equally dangerous, what was the probability of success of individual missions?

To determine cumulative probability, in this case 33%, we multiply the probabilities of individual events. So the question here is, what probability multiplied by itself 25 times equals 0.33? You can do this easily on a calculator that can raise numbers to powers. Raising 0.33 to the 1/25 power (25th root) gives an answer of 0.957, or 95.7%. Crewmen had a 95.7% chance of surviving each mission, but only a 33% chance of surviving 25 missions. Talk about good news and bad news!

Now consider how the same compounding of probabilities relates to the quality of consumer products. One way to improve quality is to reduce the number of components. If we have been using 200 components having individual reliabilities of 99.9%, we can improve overall reliability from about 82% to 98% by reducing their number by a factor of ten, i.e., to 20 components. We arrive at these results by raising 0.999 to the powers of 200 and 20, respectively. Check my math, please. Now you know one way we benefit from integrated circuits, which greatly reduce the number of components in television sets and other electronic equipment.

\* \* \*

Another example. During a visit to Puerto Rico, I was observing a roulette game. A lady walked up to the table, bought some chips and laid a bet on No. 6. The wheel was spun and No. 6 won. She picked up her winnings and left her original bet on No. 6 and it won again on the second spin. And again on the third spin. She was amazed, as well she should have been. Let's analyze it. There are 38 positions on a roulette wheel. The probability of No. 6 (or any other specified number) coming up on each spin is 1 in 38. The probability of its winning three times in succession is:

$$1/38 \times 1/38 \times 1/38 = 1/54{,}872$$

In the long run, the probability of No. 6 winning three successive times is 1 in 54,872 but it happened the first time. I too was amazed, but from a slightly different perspective. I was intrigued by any number winning three times in succession. This probability is quite different. Why? Because I didn't care which number won on the first spin. In effect, I bet all numbers on the first spin. Therefore, my probability was:

$$38/38 \times 1/38 \times 1/38 = 1444$$

Do you see the difference between the lady's case and mine? This makes statistics exciting.

\* \* \*

A final example. Have you ever wondered how many license plates your state can issue before having to change color or format? Consider the common format of three letters followed by three numbers. A license might be:

[ BNE 186 ] Assume all letters and all numerals are available. All you do is multiply out the number of ways each position can be filled:

$26 \times 26 \times 26 \times 10 \times 10 \times 10 = 17{,}576{,}000$ license plates

The small state of Hawaii has this format. It doesn't have to change very often does it?

Suppose you order a specific plate, like AAA 111. You think it is pretty unique. But it is just as probable as BNE 186 or any of the other 17,576,000 plates because in every case each letter and number can be filled in only one way, for a probability of

$1/26 \times 1/26 \times 1/26 \times 1/10 \times 1/10 \times 1/10 = 1/17{,}576{,}000$

But among all possible plates there are relatively few with identical letters and numbers. That is what makes AAA 111 special. Referring to the roulette example, can you calculate what percentage of all license plates have identical letters and/or numbers?

## CHANCE AND ODDS

Up to this point, we have only considered cases in which outcomes are either favorable or unfavorable. There are other cases in which some outcomes are inconsequential. We can rule them out by speaking in terms of "chance" and "odds." Be careful to use these terms correctly. We defined probability as the ratio of favorable outcomes to the total number of outcomes. Well, chance is the ratio of favorable outcomes to unfavorable outcomes, and odds is the reciprocal:

$$\text{Chance} = \frac{\text{Number of Favorable Outcomes}}{\text{Number of Unfavorable Outcomes}}$$

$$\text{Odds} = \frac{\text{Number of Unfavorable Outcomes}}{\text{Number of Favorable Outcomes}}$$

THE LITTLE BROWN BOOK

Assume in a game of craps that the player is rolling for a 9 to win. The house wins if a 7 comes up first. Here is how you figure the odds:

| Favorable Outcomes for Player | Unfavorable Outcomes for Player |
|---|---|
| 3 & 6 | 1 & 6 |
| 6 & 3 | 6 & 1 |
| 4 & 5 | 2 & 5 |
| 5 & 4 | 5 & 2 |
| – | 3 & 4 |
| – | 4 & 3 |

In "chance" and "odds," inconsequential rolls, such as 2 & 2, 3 & 5, etc., are eliminated from the computation because no  one wins or loses. You can see that the player has four ways to win and six ways to lose, for a 2-out-of-3 chance of winning. We normally look at it the other way around and say he has a 3-to-2 odds of losing. If the player makes a side bet on his No. 9, he will receive 3-to-2 odds if he wins, getting back his side bet plus 3/2 times the amount of the bet. On a $6 bet, the return is the $6 bet plus $9 winnings.

## LOTTERIES

State lotteries do not operate by the same rules as casinos and racetracks, which pay out most of the money bet. Lotteries are big revenue-raising measures. They pay much less than the actual odds.

In California's LOTTO games, players pick six numbers from a field of 49 numbers, and the state also draws six numbers. The betting sheet (on the back side, of course) informs players that the odds against matching three of the state's numbers is 56-to-1, for which the payout is $5 on a $1 bet. Let's analyze this. The player gets $5, of which $1 is the return of the wager, leaving winnings of $4. This is a 4-to-1 return for beating 56-to-1 odds. How about that! In a true betting situation, the return would be $56 plus return of the $1 bet. Let's talk percentages. The player wins $4 instead of $56. That's only 7.1% of what would be due on a statistical basis! The state, lottery operator and seller share 92.9%. It gets worse! They pay nothing for matching one or two numbers – the same as matching none. Would you agree that perhaps the lottery is a tax on people who are poor at math?

Try to find other situations where you can practice calculating probability or odds. Don't be like the person in the story – a joke, we hope – who reasoned that the lottery is an even-money bet because "you either win or lose."

# 4
# FROM HERE TO INFINITY

When we think of infinity, we imagine a value or a number that is without bounds, one which has no limit. We use as its symbol a figure "8" turned on its side. The interesting question is – where do we find infinite values or infinite numbers of things?

Let's begin with all the whole numbers. No matter how large a number you name, I can name a larger number. The

$$1, 2, 3, 4, 5, 6, 7, \ldots \ldots \infty$$

"limit," if we could ever reach it, is infinity. This seems pretty obvious. And it is, because we took the simplest example possible.

## SETS OF NUMBERS

Now consider some finite sets, or collections, of things. The French people are part of the European population. Therefore, there are fewer French people than there are Europeans. Likewise, there are fewer cats than animals, fewer Buicks than automobiles, fewer children than people. However, this logic does not hold for infinite sets.

The most obvious example of an infinite set is one containing all the whole numbers. What is not so obvious is that a set containing only the even numbers is also infinite. In a sense,

$$1, 2, 3, 4, 5, 6, 7, 8, 9, 10, \ldots \ldots \infty$$
$$2, \quad 4, \quad 6, \quad 8, \quad 10, \ldots \ldots \infty$$
$$3, \quad 6, \quad 9, \ldots \ldots \infty$$

there are "as many" even numbers as there are whole numbers. This property was once thought to be a contradiction, but today it is used as a definition of infinity. The first set is twice as large as the second set, yet each is infinite in size. Similarly, we could have chosen every third number, etc.

What happens when we deal with decimal numbers? Do you think there is a finite or infinite number of decimal numbers between 1 and 2, or between 1051 and 1052? The answer – an infinite number because our decimal system allows us to create values to an unlimited number of decimal places. This is an example of an infinite number of infinite sets.

We also encounter the concept of infinity in geometry. Students know that there is an infinite number of points on a line. Moreover, between any two points there is an infinite number of other points. The explanation, of course, is that a point is dimensionless; it occupies no space. A pair of points defines a line segment. The segment can be infinitely short. In this context, we can find endless infinities on the head of a pin.

## THE REAL WORLD

Infinity is a necessary mathematical notion, but can we find an infinity of anything around us? Is the number of grains of sand infinite? Definitely not. We could count them if we had the time and patience. Someone once estimated the number of grains of sand on the beach at Coney Island, New York to be about $10^{20}$, i.e., 100,000,000,000,000,000,000, or 100 billion billion. You might want to extend this to all the beaches and deserts in the world. But the number is still countable, isn't it?

An article in a distinguished scientific journal once estimated that the number of snow crystals required to form the Ice Age was one billion billion or $10^{18}$, about the same as the number of seconds that have transpired since the Big Bang. The total number of words ever spoken is about $10^{16}$, somewhat fewer than the number of words printed since the invention of the printing press.

If these immense numbers sound like mathematical oddities, refer to the opening chapter on our Space Odyssey. A light-year is $6 \times 10^{12}$ miles and the most distant known galaxy is $6 \times 10^{22}$ miles from earth. You see, astronomers work with these sizes of numbers every day. Want a bigger number? Convert the distance to inches or millimeters.

The reason for all these examples is to emphasize the fact that no matter how large a value we find, it will be a finite one. Anyone claiming an infinite size or quantity in the real world is not a mathematician. An interesting possible exception is the extent of the universe. Is there any limit to the expanse through which debris from the Big Bang is expanding? Cosmologists tell us that most scientific evidence indicates the universe will expand forever. Does this in some way imply that space is infinite?

# 5
# THE RIGHT SIZE

About sixty years ago a British biologist, Professor J.B.S. Haldane, wrote an essay titled On Being the Right Size. It answers many questions about the animal world around us.

## FALLING OFF A BUILDING

Mice and other tiny animals can fall without harm from the top of a skyscraper. A rat would be killed, although it would survive a fall from an eleven-story building. Humans fall only a couple of floors and are killed. Why these differences?

The answer lies in the way changes in size affect volume and surface area and the fact that the resistance which air presents to a falling body is proportional to surface area. Let's make an observation about volume and surface area.

$$\text{Volume} = L \times W \times D$$
$$\text{Surface Area} = 2LW + 2LD + 2WD$$

If we reduce all dimensions to 1/10 size, we see that:

$$\text{Volume (1/10 scale)} = \frac{L}{10} \times \frac{W}{10} \times \frac{D}{10} = \frac{\text{Original Volume}}{1000}$$

$$\text{Surface Area (1/10 scale)} = 2 \times \frac{L}{10} \times \frac{W}{10} + 2 \times \frac{L}{10} \times \frac{D}{10} + 2 \times \frac{W}{10} \times \frac{D}{10}$$
$$= \frac{\text{Original Surface Area}}{100}$$

So, if a body reduces to 1/10 its original size, its volume

reduces to 1/1000, but its surface area to only 1/100. This is also true for spheres, whose volume and surface area are proportional to the cube and the square of the radius, respectively. This tells us that volume, and therefore weight, decreases ten-times faster than surface area. Since weight decreases so much faster than surface area, air resistance becomes more effective in breaking the fall. The greater the reduction in volume, the greater the effect of air resistance. Therefore, the smaller the animal, the farther it can fall without injury. Archimedes knew this in 300 B.C.

## WALKING ON WATER

We have shown that insects should have no fear of falling. They can even walk on a ceiling. And they can walk on water. But watch out little insect! Don't break the surface tension of the water, i.e., don't get wet. Because if you do, an amount of water many times your weight will adhere to your body, and you won't be able to escape. As Professor Haldane said, "An insect going to a puddle of water to get a drink is in as great a danger as a man leaning over a precipice in search of food."

Larger animals are not burdened by water on their bodies. Why? Our prior theory applies in reverse. As a creature becomes larger, weight increases faster than surface area. Water that clings to the slowly increasing surface area becomes an ever smaller percentage of body weight. When you step out of the shower, you carry an added weight of less than one pound.

## HOW TALL IS TOO TALL?

Suppose a human giant existed. What might be its weakness? It would be a structural weakness. If a person's three dimensions increased by a factor of ten times, weight would increase by 1000 times. But the legbone, whose strength is proportional to cross-sectional area, would increase by only 100 times. Might such a giant easily break a leg while walking? Also the giant's heart would have to pump blood to greater heights causing higher blood pressure and requiring stronger blood vessels.

As larger species of animals evolved, their limbs became relatively shorter and thicker and their skeletons bulkier and heavier. Bones make up about 8% of the weight of a mouse, 13% of a dog and 17% of a human. Evolution avoided the defects of our proposed giant. Larger animals are also more clumsy. Compare the agility of a cat to a human, to an elephant. On the other hand, in the evolutionary process, it is interesting to note that there is less skeletal difference between a porpoise and a whale. Why? Because the buoyancy of their water environment neutralizes the effects of weight differences.

The famous astronomer and scientist, Galileo Galilei, wrote about the self-limiting heights of ships and buildings, even trees. He suggested 300 feet as the limit for trees.

## KEEPING WARM

An advantage of our large size is that it is easier to keep warm. When at rest, all warm-blooded animals lose essentially the same amount of heat per unit of skin area. Therefore, they need a food supply proportional to their surface area – not their volume. Since surface area decreases only 1/10$^{th}$ as fast as volume, small animals require relatively more food than large animals. Five thousand mice weigh as much as a human, but their combined food requirement is ten times as great. A mouse eats about a quarter of its weight in food each day, primarily to keep warm. That's why small animals cannot survive in polar regions and why insects die during the winter. Some animals developed a fur coat for protection. This is nature's way of providing a better balance between volume and surface area. Humans do not have fur coats. Now you know why.

## WHY CAN'T WE FLY?

As birds become larger they develop more power, but the speed needed to become airborne also increases. The problem is that the rate at which power increases is lower than the rate at which the required speed increases. A limit is soon reached where a bird is too large to fly. The ostrich is a good example. It needs a speed of 100 MPH to become airborne, but it cannot run that fast.

In films of man's early attempts to fly, you see men running as they flap large wings attached to their arms. They were emulating birds, but they did not understand the physics of the problem. They could never run fast enough to lift off.

## OPTIMUM SIZE

Each species of animal has an optimum size. Any large increase in size inevitably requires a change in form. Large animals are more complicated than small ones. They also move slower and live longer. However, they are not larger because they are more complex. Rather, they are more complex because they are larger. Might we apply the same observations to governments and corporations?

## VOYAGING INTO SPACE

If sometime in the next century we manage to send astronauts on a voyage to Mars, we will need to have solved the problem of prolonged weightlessness.

The human body is optimized to function in earth's gravitational field. When gravity is greatly reduced for long periods, we require less food, less blood, and less lung power. With no gravity to resist pumping, our heart grows lazy and shrinks in size, and our muscles weaken. For full details, refer to Mission to Mars.

In going to Mars we will experience for as long as a year in each direction an environment to which our bodies are not optimized. Never in millions of years of evolution were our ancestors exposed to anything like the weightlessness of space travel. Like a human falling off a ten-story building or a mouse stranded at the South Pole, our astronauts will perish if we don't provide adequate protection.

# 6
# PLANETS IN MOTION

Greek writings dating from as early as Aristotle in the 4$^{th}$ century B.C. reveal knowledge that the earth is round. (Why do we insist on saying "round" when we mean spherical?) Eratosthenes even calculated the circumference of a great circle that passed through the cities of Syene (now Aswan) and Alexandria. But there was no agreement on how the earth moved or even if it moved.

In the 2$^{nd}$ century A.D. the Greek geometer, Claudius Ptolemy, published convincing theories that the earth was fixed in space and that the sun and planets traveled around it. Although competing theories held that the reverse was true, Ptolemy's concept of the universe was generally accepted for more than a thousand years.

Let's pick up the story in the 1500s when Europe was engulfed in a great dispute over the relative motion of our sun and its planets. The 1500s may seem like long ago. But was it? At 25 years per generation, this was about 18 generations ago. Are your great grandparents still alive? If so, they represent the third generation back. Maybe 18 generations is not so far back, after all, for so basic a fact to have remained unresolved.

The argument as to whether Earth revolved around the sun (heliocentric theory) or the sun around Earth (geocentric theory) involved three great mathematicians.

| | |
|---|---|
| Nicolaus Copernicus | (1473-1543) |
| Johann Kepler | (1571-1630) |
| Galileo Galilei | (1564-1642) |

## NICOLAUS COPERNICUS

Nicolaus Copernicus was a Polish scholar and priest who had studied law and medicine in Italy. His work included currency reform and advising the Pope on calendar reform. He was also an astronomer. Using only geometry and the siderial periods of revolution, he calculated the distances between the sun and the six known planets. He is credited with reviving and supporting the theory that Earth revolves around the sun. Although his first nonmathematical sketches may have appeared before the age of forty, he made no formal publication for fear of reprisal by the Church, which was steadfast in its teaching that Earth was the center of the universe. Ironically, it was no longer so much a matter of doctrine as it was authority. Yet this matter would not be resolved until many years later. In 1543, near the age of seventy, Copernicus decided to publish his mathematical description of the heavens anyway in what he called, *The Revolution of the Heavenly Orbs*. The first copy was put into his hands as he lay dying.

## JOHANN KEPLER

Johann Kepler was a German mathematician who worked in Czechoslovakia and Austria. He supported the heliocentric theory and worked passionately to identify the path of planetary motion. The Greeks knew long ago that planets do not travel in simple circles, but they did not attempt to discover the actual path. This, Kepler set out to do. Investigation of an oval path set him in the right direction. After years of ardent work, he identified the path as an ellipse.

An ellipse is a special kind of oval. It is generated as illustrated in the following drawing. On the left, the sum of the distances from its foci $F$ is a constant. However, it is not so much the path that interests us, as it is the motion of a planet

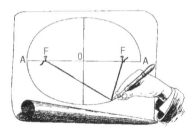 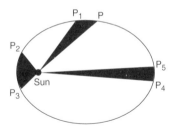

in that path. As a planet moves in its elliptical orbit, with the sun at one of the foci, it sweeps out equal areas in equal times. Since the shaded areas in the drawing on the right are all equal, the planet must travel from $P$ to $P_1$, $P_2$ to $P_3$ and $P_4$ to $P_5$ in equal times. But these distances are not equal. This means that our huge planets – and also the stars – are continuously changing speed, correct? How amazing! All these things confirmed Kepler's observations. He had succeeded in turning Copernicus' general description of the sun and planets into a precise mathematical formula. And a very simple one, at that.

Let us pause and note that when the foci are relatively far apart, as with the distant planets Pluto and Neptune, the orbit is considerably elongated. Conversely, planets closer to the sun have more circular orbits because their foci are closer together. In fact, a circle is a special case in which the two foci are coincident.

Since the Church was indifferent to orbital information, this episode was not affected by the controversy over heliocentric theory. But his contemporary, Galileo, was not so fortunate.

## GALILEO GALILEI

Galileo was born in Italy in the same year that William Shakespeare was born in England. He had interests in astronomy by virtue of his work with telescopes and he described his astronomical observations in his famous book, *The*

*Galileo (1564-1642), Italian mathematician, astronomer and physicist.*

*Starry Messenger.* He even made the first observations of Jupiter's four largest moons, thus exploding forever the notion that all heavenly bodies orbit the earth. Galileo became convinced that Copernicus had been correct. He began publishing these ideas, knowing full well that they were in opposition to the teachings of his Church. But these were the days of the Inquisition! In 1616 Galileo was ordered not to uphold or even defend Copernican theory.

Galileo would not be intimidated. Confident that he was correct, he published in Florence in 1632 his *Dialogues of the Great World Systems* in which one speaker objects to Copernican theory and two others answer the objections. For this he was brought to trial in Rome, despite his having collected no fewer than four imprimaturs for his book and despite the fact that it was immediately acclaimed as a masterpiece.

This was no minor trial. There were ten judges, all cardinals, including a brother and a nephew of Pope Urban VIII who in his younger days had written a sonnet of compliments to Galileo on his work in astronomy. Galileo agreed to recant, although legend has it that when he arose from kneeling before his inquisitors he murmured in Italian, "Even so, it does move." He was put under house arrest in Sienna. Twice over the years he was threatened with torture.

Modern historians say that the Church acknowledged that Galileo was correct, that its own astronomers were saying the same thing, that it wanted to break the news gently to its followers and that Galileo was brought to trial because he would not suppress his ideas in the meantime. Not until 1992 after a 13-year investigation did the Church formally acknowledge that it had wrongly opposed Galileo, thus ending one of the Inquisition's most infamous wrongs.

The effect of Galileo's trial and imprisonment was to stifle the scientific tradition of the Mediterranean region. The scientific revolution moved to northern Europe. Galileo died, still under house arrest, in 1642. On Christmas Day of the same year, Isaac Newton was born in England.

## ISAAC NEWTON

One of Sir Isaac Newton's greatest accomplishments was his discovery of universal gravitation – the "glue" that holds our universe together. He included this idea in his famous book, The Principles of Natural Philosophy, or *Principia*, for short.

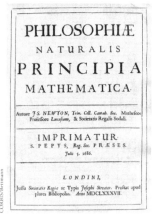

Perhaps the most revolutionary part of his concept was not so much that the sun pulls Earth, but that Earth also pulls the sun – and pulls equally hard. And that the universe is full of moving bodies and they all pull each other, down to the smallest grain of sand. Do a boy and girl echo Newton's law of gravity when they say they are falling for each other, or are attracted to each other?

Newton's *Principia* vindicated the "heresies" of Copernicus and Galileo. On the basis of Newton's work, later astronomers were able to predict the existence of Neptune and Pluto long before we had instruments to observe them. Knowledge of gravity and Newton's laws of motion enabled us to navigate to the moon, to explore distant planets with space probes, and now to plan a journey to Mars.

# 7
# LET'S TALK PERCENTAGES

Let us begin by examining how businesses present prices to the public. When banks advertise for deposits, they don't say "We pay 5.5¢ per year for each dollar deposited," or "We pay $5.50 for every $100." No, they say "We pay 5.5% per year." Regardless of the amount deposited, you multiply it by 5.5/100, or 0.055, to determine the first year's interest. The bank's message is short and complete. One statement covers all deposits. This is an example of good communication through the use of percentages.

We find a similar situation at clothing stores. When conducting a sale, they don't advertise they are cutting the price of their $62.99 sweaters to $56.69, their $28.48 sweaters to $25.63, etc. Of course not; they advertise "10% off on all sweaters." Another example of efficient communication.

In other areas we are not so fortunate. The financial world (of all places) and the news media have a long way to go in adopting percentages. "Stock A closed today at 93 3/4, down 1 1/4, and Stock B closed at 12 1/2, down 1/2." Listen carefully and you will hear many announcers emphasize the stock with the larger dollar decline. How ridiculous. This is merely raw data! If we want an accurate comparison of the actions of the two stocks, we must compare their percent changes with respect to their prior closing prices:

Stock A: $-\$1.25/(\$93.75 + \$1.25) \times 100 = -1.3\%$
Stock B: $-\$0.50/(\$12.50 + \$0.50) \times 100 = -3.8\%$

Isn't that interesting? Stock A lost 1.3% of its value, while Stock B lost 3.8%. Stock B sustained the greater loss in value. And by a factor of almost 3-to-1. Yet Stock A was emphasized in the news. You can find examples like this every day. Practice mental calculations while watching the financial news. I saw an experienced commentator emphasize a 5 point loss in an index value. It turned out the value was 1550. I didn't need a calculator to estimate the loss as about 1/3 of 1% – like a $35 stock losing 1/8. It's sad, but true.

Even when percentages are used, you must be alert. Here is a notorious example of misuse or, worse yet, disinformation. Your state is increasing the sales tax from 6% to 6.6%. It announces that the tax is going up "0.6%." Wrong! The tax is going up 10%. The state will collect 10% more tax dollars on each sale, will it not? If they wish to avoid this dire message, the proper statement is "a 0.6 percentage point increase" in sales tax. We must expand our use of percentages, but we must do it correctly.

Finally, learn to be an aware listener and reader because the media report many falsehoods. I think this is best illustrated by continual reports during recent years that there are 2.5 million homeless people in our country. You should immediately reject this number as false. Not erroneous, but false! Why? It is very simple. You know that our total population is about 250 million, so 2.5 million is 1%. You don't believe for a moment, do you, that one of every 100 men, women and children in our country is living on the streets? Public debate suffers due to distortions of such readily verifiable facts. In general, be wary of any data presented in terms of raw numbers, especially where "causes" are involved. It is much easier to fool people with raw numbers than with percentages.

Another form of deception is the omission of essential facts. When someone reports that we spend "40% less" on civilian R & D than Germany or Japan, ask immediately whether this is 40% fewer dollars or 40% less of our GNP? If it's the latter – and it is – the U.S. actually spends more money, because our GNP is so much greater than that of any other country. Sadly, the writer wants readers to infer that we spend less money.

Finally, be cautious with any graph whose scale does not extend down to zero. This zero suppression allows expansion of the top of the curve in order to dramatize short-term variations. If the base of the graph were zero, recent variations would be hardly noticeable. The name of the game is communication. Do you agree that a graph of percent change would be more informative?

Now, if percentages are so valuable for evaluating current data, won't they serve equally well to track data over long periods of time? The answer is a resounding, yes.

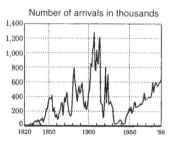
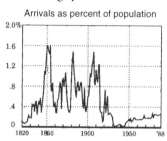

Here are two graphs showing immigration statistics for the past 170 years. The graph on the left charts the number of immigrants each year. It shows that immigration increased on almost a straight-line basis from about the time of the Civil War (1865) until 1920, then declined precipitously until 1940, and is now on a sharp rise again. The graph on the right charts immigration as a percent of each year's population. It shows that immigration cycled between 0.4 and 1.5% from about 1865 to 1920, then declined steeply until 1940 and since 1950 has held fairly steady at about 0.2% of population.

Which analysis is correct? Obviously, they both are. They present the same information. They just do it in a different manner. Which is most useful? It depends on one's motive. An advocate of immigration quotas would use the raw data. Anyone with a business interest would use the percentage data. If you are a city planner preparing budgets for schools and community services, you need to know by what percent you must increase your budget each year to accommodate the special needs of immigrants. If your community is typical of the entire country, the percentage chart indicates you need about a 0.2% annual increase. It also suggests there is little likelihood of a significant change anytime soon.

## LOGARITHMIC SCALE

Have you studied logarithms? If not, it won't matter. We can get around it. Logarithms appear widely in scientific work. Here is an elegant way we use them to improve our graphics.

In the late 1500s, John Napier was studying the history of decimal notation when he noted an important repetition in our counting system:

|   |   |   |   |   |   |
|---|---|---|---|---|---|
| 1 | 2 | 3 | 4 | 5........ | 10 |
| 10 | 20 | 30 | 40 | 50....... | 100 |
| 100 | 200 | 300 | 400 | 500...... | 1000 |

He reasoned that if he made 1, 10, 100, 1000, . . . appear at equal intervals on a scale, then 2, 20, 200, . . . ; 3, 30, 300 . . .; etc., must also appear at equal intervals. What Napier had discovered was the concept of an equal-percentage, or logarithmic, scale. But where should he locate the values 2 through 9? He laid aside all other work and spent the next 25 years of his life developing his table of logarithms that tell us where to place intermediate values:

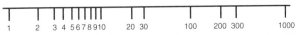

Here is why we call this an equal-percentage scale. Choose any value and repeatedly multiply it by a constant, e.g., 1, 2, 4, 8, 16, 32; or 10, 100, 1000. The values in each sequence are equally spaced, meaning that equal percentage changes (in this example 100% and 900% respectively) produce equal increments on the scale. You shall soon learn how advantageous this can be. In contrast, a linear scale produces equal increments for equal additions to values. These additions are high percentages at low values and low percentages at high values – not an ideal situation for many analyses.

Let's put a 'log' scale on the vertical axis of a graph and plot value vs. time. We show just one cycle (decade) of data above and below an initial or a reference value. Note that a log scale does not have a zero. That's just fine. Since we can extend the scale as many decades as needed below the reference, we can accurately plot extremely small values.

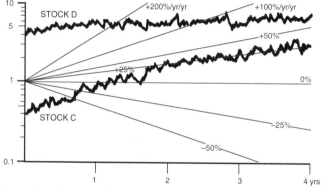

Now for the payoff. If we plot a value that changes at the same rate year after year, the points will all lie on a straight line, e.g., 100%/yr/yr. We can overlay the graph with a family of equal-percentage lines which we can visually slide up, down and across the graph to overlay and interpret our data.

By examining its slope, we see that Stock C increased about 100% each year (parallel to the 100%/yr/yr line) for the first two years and then leveled off to about 25%/yr/yr. This is typical of young companies. They cannot sustain a 100% growth rate. Stock D increased about 5%/yr/yr. This is characteristic of mature companies. The beauty of the logarithmic scale is that it enables us to evaluate data at a glance and to do it reliably.

The chart below uses a log scale to present the history of the Dow Jones Industrial Average for the past 70 years. Observe what happened between 1940 and 1965. It grew at almost a constant rate from 100 to 800. A 9.5%/yr/yr line would overlay this segment perfectly. Then it held pretty flat until the early 1980s when it took off at a 12.5% growth rate which continued for the rest of the period. We would have a devil of a time making this summary if the data were plotted on a linear scale.

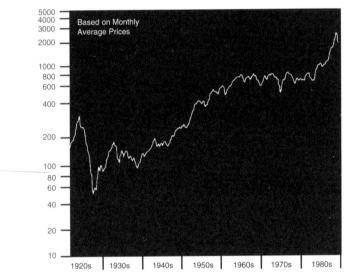

But we never get something for nothing. All things come at a cost. There is always a tradeoff. The tradeoff here is that the log scale smoothes the curve. Short-term variability is obscured because short-term changes are usually small in terms of percent, so the logarithmic scale usually works best for long-term charts. But we must always think percentages. Let there be no doubt about that.

# 8
# A NUMBER IS A NUMBER, OR IS IT?

What is your definition of a number? It probably relates to counting. And this, of course, is where the practice of numeration began. But it has evolved over the centuries into a much more interesting and complex subject. Our story begins with the need to count – probably the first mathematical step taken by the ancients:

CARDINAL NUMBERS:   1, 2, 3, 4, 5, 6, 7, . . .

These "whole" or "natural" numbers enabled our early ancestors to keep track of their sheep, pigs, baskets of grain, and other possessions. Later, they used these same numbers to define order or sequence:

ORDINAL NUMBERS:   $1^{st}$, $2^{nd}$, $3^{rd}$, $4^{th}$, $5^{th}$, . . .

Given more time, the need evolved to measure distance, so the numbers took on an additional meaning of magnitude. But there had to be a starting point, so they created zero. And things could go backward as well as forward, so they created negative numbers:

INTEGERS:   . . . –4  –3  –2  –1  0  1  2  3  4 . . .

At this point we have a very ordered system for forward and backward progress. It also accommodates addition and subtraction. Addition means "go forward." Subtraction means "go backward." We can also multiply, because multiplication is a form of addition. Correct? "3 times 5" means "5 + 5 + 5."

Each of these operations always results in a whole number.

But what about division? Division is a form of subtraction. "10 ÷ 5 = 2" means we subtract 5 from 10 two times and have nothing left over. But when we divide 11 by 5, we have 1 left over. Division produces a fraction more often than not. They had no way of defining a part of a basket of grain or a part of their unit of distance. So they created a set of fractional numbers to fill the gaps between the integers:

RATIONAL NUMBERS: e.g., 1/2, 2/3, 25/32, 243/308

Rational numbers result from dividing one whole number by another. We often express them in their alternative decimal form. Incidentally, integers are included because they have a divisor of one. Mathematicians were pleased. They had filled the gap between each integer with an infinity of fractional numbers. They thought they had done it all. Everything went along swimmingly. That is, until Pythagoras came on the scene.

We are in Greece in the 5$^{th}$ century B.C. Pythagoras was playing around with right triangles when he discovered that the length of the hypotenuse equaled the square root of the sum of the squares of the other two sides. In the simplest case, where the sides are of length 1, the length of the hypotenuse is equal to $\sqrt{2}$. Now what? Can we locate $\sqrt{2}$ among the rational numbers? No, because it cannot be expressed as a fraction. What a paradox! They had an infinity of numbers between 1 and 2, and $\sqrt{2}$ was not among them. They soon realized that they had discovered a new kind of number:

IRRATIONAL NUMBERS: e.g., $\sqrt{2}$, $5.79^{1.3}$, $5\sqrt{143}$

At this news, the Greek philosophers celebrated by sacrificing an integer number of oxen, namely 100. And momentous it was, because they had discovered a linkage between numbers and positions. We are still impressed by their accomplishment. The Pythagorean Theorem is said to be the most-often-proved theorem in all of mathematics – 370 known proofs including one by President James Garfield. It was suggested as an item to place aboard our spacecraft as proof of our intelligence, in event of interception by voyagers from another planet. One would think that getting a spacecraft up should be sufficient proof!

You can see that there is an infinity of irrational numbers, as well as rational numbers, between each integer. In preparation for the next step, we must pause and note that rational and irrational numbers can solve algebraic equations, such as $ax^2 + bx = c$, where $a$, $b$ and $c$ are rational numbers. As if we didn't have enough numbers already, we later found an infinity of numbers that cannot solve algebraic equations:

**TRANSCENDENTAL NUMBERS:**   e.g., π, e, log 2, ln 85

Their existence was proved by Joseph Liouville, a French mathematician, in the early 19th century. The term "transcendental" was coined when these numbers were thought to be very rare. Then Alexander Gelfond proved a hundred years later that $a^b$, where $a$ is any rational number other than 0 or 1, and $b$ is any irrational number, is transcendental. That's a bunch!

Now, let's return to an earlier era when mathematicians were confronted with equations whose solutions were the square root of negative numbers – for example $x^2 = -1$, for which $x = \pm\sqrt{-1}$. René Descartes, the 17th-century French mathematician and coinventor of our X-Y coordinate system (hence the name Cartesian coordinates), rejected the possibility of an equation having such an "impossible" root. But he was later proved wrong. Now we had another kind of number for which we created the symbol '$i$:'

**IMAGINARY NUMBERS:**   e.g., $\sqrt{-9} = \pm 3\sqrt{-1} = \pm 3i$

An imaginary number is unlike any number we have seen so far because it is neither greater nor smaller than any other number. It cannot be compared with the others which we shall now call "real" numbers. It was the ingenuity of Karl Friedrich Gauss that gave us a practical use for imaginary numbers:

COMPLEX NUMBERS:   e.g., 2 + 3$i$; 3 + 2$i$

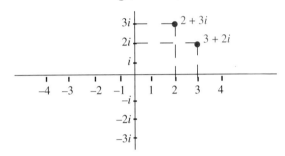

Gauss invented the scheme of defining X-Y coordinates by using a real number to designate the X-axis coordinate and an imaginary number for the Y-axis. Instead of identifying a location on a plane as $x = 2$, $y = 3$, for example, he identified it as 2 + 3$i$. By so doing, he expanded the use of numbers from defining a point on a line to defining a point on a plane. The idea was later extended to space by adding a '$j$' component to define a point on the Z-axis perpendicular to a plane. For example, 2 + 3$i$ + 4$j$ is a point in space 4 units above the point 2 + 3$i$.

## CLASSES AND SETS OF NUMBERS

Returning to the cardinal numbers, we can extract various classes or sets from them. Some are oddities, at best. But mathematicians enjoy searching for them.

ODD & EVEN NUMBERS:   1, 3, 5, 7, 9, . . .
                      2, 4, 6, 8, 10, . . .

A Number is a Number, | 41
or is It?

The even numbers are evenly divisible by 2; odd numbers are not. There is an infinite number of each.

PRIME NUMBERS: 1, 2, 3, 5, 7, 11, 13, 17, 19, 23, . . .

A prime number is evenly divisible only by itself and 1. Euclid, who gave us geometry, proved there also is an infinity of these. The highest known prime is $2^{1,398,269} - 1$, containing 420,921 digits. It was discovered by a supercomputer in 1992. For us in the real world, Gauss and Legendre's Prime Number Theorem says that the proportion of numbers below any value $n$ that are primes is approximately equal to the reciprocal of the natural logarithm (ln) of $n$, when $n$ is large. Punch up the ratio 1/(ln 10,000) on your calculator. Now you know that about 11% of the whole numbers between 1 and 10,000 are prime numbers, or about 22% of the odd numbers because, except for 2, the even numbers are not prime. The rest are called composite numbers.

PERFECT NUMBERS: 1, 6, 28, 496, 8128, .

A perfect number is the sum of its divisors, e.g., 6 = 1 + 2 + 3, or 28 = 1 + 2 + 4 + 7 + 14. The highest currently known is (22,859,433 − 1) x 22,859,433, containing 517,430 digits. How about this as an example of how unimaginably immense a number can be and still be finite. For more math trivia, see the Guiness Book of Records.

Euclid told us that if $2^{n-1}$ is a prime number, $2^{n-1} \times (2n - 1)$ is a perfect number. Try it for $n$ = 1, 2, 3, 5. It works. In the 17$^{th}$ century the Swiss mathematician, Leonard Euler, proved that all even perfect numbers are of this form.

PYTHAGOREAN TRIPLETS: e.g., 3, 4, 5; 5, 12, 13; 7, 24, 25

Solutions to a right triangle where all sides are of integer length, e.g., $3^2 + 4^2 = 5^2$, are called Pythagorean Triplets. Interestingly enough, no one has ever found a solution in positive integers for the equation $a^N + b^N = c^N$, when $N$ is an integer greater than 2. Nor has anyone ever proved or disproved the existence of a solution.

**FIBONACCI SEQUENCE: 0, 1, 1, 2, 3, 5, 8, 13, 21, . . .**

This is a sequence in which each number (except for the first two) is the sum of the two preceding numbers. It takes the surname of Leonardo of Pisa, a 13th century Italian merchant who upon returning from the Orient helped introduce Arabic numerals to his homeland. It resulted from solving the problem of how many rabbits there are in successive generations. It also appears in other places. In a sunflower, for example, the numbers of spirals to the right and to the left are generally adjacent numbers in the Fibonacci sequence. It also appears in the symmetry of nautilus shells. Finally, statisticians find this sequence in the cyclical characteristics of commodity prices, and they use it to predict price trends.

You can easily find other groupings with particular characteristics. Good examples are the binary and power-of-three numbers used on pgs. 9-10.

# 9
# SCIENTIFIC NOTATION

You have seen plenty of scientific notation in earlier chapters, with more to come. And why not? Imagine my telling you that about 1,000,000,000,000,000,000 seconds have elapsed since the Big Bang. Now, quickly, how much larger is this number than the 10,000,000,000,000,000 words ever spoken? How long did it take to answer? But if I give you the first number as $10^{18}$ and the second as $10^{16}$, you respond immediately with $10^2$, or 100-times. That's one reason we use scientific notation.

We express a number as the product of the number itself with the decimal point repositioned behind the first non-zero digit and a power of ten. To determine the power, count the number of positions you move the decimal point, + for backward and – for forward. For example:

$$69417 = 6.9417 \times 10^4$$
$$0.0069417 = 6.9417 \times 10^{-3}$$

My calculator gives the answer in this form whenever a computation overflows the display. For ease of typing, we occasionally see the numbers expressed as 6.9417 E+04 and 6.9417 E –03.

Scientific notation has many advantages. We save time in comparing values. We reduce our chance of error. We save space. We avoid having to memorize all the ratio prefixes. There are not enough of them anyway. Perhaps most importantly, we can multiply and divide mentally by using the law of exponents which tells us that $10^M \times 10^N = 10^{(M+N)}$ and conversely, $10^M \div 10^N = 10^{(M-N)}$. When the answer comes up $10^0$, we recall that its value is 1.

It is common practice in engineering and scientific notation to use ratio prefixes to simplify numerical quantities, e.g., 1 kHz (1000 Hertz), 1 MByte (1,000,000 Bytes), 1 μsec (0.000001 second). The Greek alphabet is also commonly used in mathematical notation along with the italicized Roman alphabet, e.g., $\lambda = c/f$ (wavelength of sound = speed of sound/sound frequency).

Use scientific notation wherever it is practical. As for myself, I'm going to relax and read a chapter or two of *The Tales of $1.001 \times 10^3$ Nights*.

| Ratio Prefixes | | | | | |
|---|---|---|---|---|---|
| Symbol | Prefix | Ratio | Symbol | Prefix | Ratio |
| T | tera | $10^{12}$ | c | centi | $10^{-2}$ |
| G | giga | $10^9$ | m | milli | $10^{-3}$ |
| M | mega | $10^6$ | μ | micro | $10^{-6}$ |
| k | kilo | $10^3$ | n | nano | $10^{-9}$ |
| h | hecto | $10^2$ | p | pico | $10^{-12}$ |
| da | deca | 10 | f | femto | $10^{-15}$ |
| d | deci | $10^{-1}$ | a | atto | $10^{-18}$ |

| The Greek Alphabet | | | | | |
|---|---|---|---|---|---|
| A | α | alpha | N | ν | nu |
| B | β | beta | Ξ | ξ | xi |
| Γ | γ | gamma | O | o | omicron |
| Δ | δ | delta | Π | π | pi |
| E | ε | epsilon | P | ρ | rho |
| Z | ζ | zeta | Σ | σ | sigma |
| H | η | eta | T | τ | tau |
| Θ | θ | theta | Υ | υ | upsilon |
| I | ι | iota | Φ | φ | phi |
| K | κ | kappa | X | χ | chi |
| Λ | λ | lambda | Ψ | ψ | psi |
| M | μ | mu | Ω | ω | omega |

# 10
# MEAN, MEDIAN, MODE

We are continually called upon to calculate the central tendency of data related to sports, economics, scholastic achievement and social statistics. We make so much use of what we call the "average" that we often overlook the fact that other measures are more representative of some data.

In past centuries when sea voyages were much more hazardous than today, deck cargo was thrown overboard during storms to make the ships more stable. Risk was shared by everyone who had cargo aboard. Compensation for lost cargo became known as "havaria." From this Latin word, we derive our modern word "average."

The idea of an average is familiar to everyone. We have batting and bowling averages, grade-point averages, average wage and height, etc. You know that we compute averages by adding all individual values, $a_N$, and then dividing by their total number, $N$:

$$\text{Arithmetic Mean} = \frac{a_1 + a_2 + a_3 + a_4 + \ldots a_N}{N}$$

In mathematical terminology, the average is called the "mean" or, more properly, the arithmetic mean. It is the measure of central tendency that best describes what we commonly refer to as the average.

Let's compute the mean weight of the boys in our class. And while we are at it, let's record their individual weights in rank order (from lowest to highest):

```
                    71 - 84 - 85 - 85 - 86 - 88 - 92 - 93 - 94 - 94 - 94 - 98 - 109
                                              |    |                   |_____|
                                            Mean  |                            |
                                               Median                         Mode
```

Adding the weights of all 13 students, and dividing by 13, we see that their mean weight (to the nearest pound) is 90 lbs. Alternatively, we could say that the "median" weight is 92 lbs, meaning that as many boys weigh less than 92 lbs as weigh more than 92 lbs. Or, we could say that the "mode" (from the French, *a la mode*) is 94 lbs, meaning that more boys weigh 94 lbs than weigh any other single weight. We now have three different descriptors for central tendency. Which should we report? Which best describes the average weight? We can approach this question more intelligently by asking if any of the measures is significantly less reliable than the others.

## MEAN

Study the data. You may say to yourself, "Gee, we weighed everyone right after lunch, and I saw one of those 94-lb boys eat a big banana. What if he had given it to the 84-lb boy to eat?" That is an excellent question. What would have happened? The boy who gave away the banana may have weighed only 93 lbs, and the one who ate it may have weighed 85 lbs. This innocent action would have lowered the mode from 94 lbs all the way down to 85 lbs, while the mean and median would not have changed at all! We had better not report the mode because it is too susceptible to minor variations in the data. This is often the case.

In practice, we usually prefer the mean. More people are familiar with it. And when the distribution is reasonably symmetrical (as in our example), the mean is least affected by minor variations in the data.

There is another reason for preferring the mean. Since it is the only measure that is derived from a formula, it is the only one on which you can perform subsequent computations. For example, if you read that the average value of the homes in your community is $155,000, you can multiply by the total number of homes and thereby estimate the total residential real estate value in your community. Do you see that you could not do this with the median or the mode? The mean is also the best choice if we want to use our data to infer knowledge about a larger population. For example, if 15 students in a class of 100 take an examination, their mean score is the most reliable estimate of what the average would have been if everyone had taken the test.

## MEDIAN

Observe that the median value is much easier to obtain. Why? Because we don't have to perform any addition and division. We simply identify the middle value. That's all there is to it. In fact, we can go one step further and note that we don't even have to weigh all the boys in our class. For sure, we can look at them and dispense with weighing the 71 and 109 pounders. They don't affect the median value, do they? Suppose you are surveying wages in a local factory. If you report the median wage, you won't have to request the salaries of the top executives, will you? No. You know they are the highest paid, so you pair them off with an equal number of the lowest paid employees, and drop all of them from your survey. Perhaps we have gained some insight into why government statistics dealing with hourly wages, annual income, wealth, home value, etc. are reported as median values.

The median leads us to another option. In large populations, the median is often referred to as the "$50^{th}$ percentile," meaning that it is greater than 50% of all values in the population. Likewise, a value at the $95^{th}$ percentile is greater than 95% of all the values, etc. This leads to grading SAT scores, cholesterol count, blood pressure, etc. in percentiles, in order to tell people where they stand in their population group. Do you see that a percentile ranking conveys more information than a raw test score or a medical measurement?

## MODE

By this time, you are probably wondering if we ever use the mode. So far, we have discussed only numeric data, such as dollars, weight, test scores, etc. But when dealing with nominal values, such as characteristics, habits, actions and other attributes having nothing to do with numbers, we use the mode. And we do it without ever thinking of it as a mathematical operation. "White is the most popular automobile color" means that more people buy a white car than buy any other individual color. White is the mode, or the modal color. There is no such thing as an average color, is there? And the median makes no sense either.

## WEIGHTED AVERAGE

The arithmetic average is valid only if the individual values contribute to the result on a one-for-one basis. One person, one vote, so to speak.

There are exceptions to this rule. A good example is the inflation index of grocery prices. Would you average the unit prices of one quart of milk, one bag of potato chips, one pound of salt, one pound of coffee, one head of lettuce, etc.? I hope not! Your family may actually use five quarts of milk every week, one pound of coffee every two months, one pound of salt every year, etc. Common sense tells us to average the cost of the proportionate amount of each item used every week. This is called a "weighted" average. The government uses it to compute the Consumer Price Index.

## MOVING AVERAGE

The financial community employs a graph of moving average to smooth random day-to-day behavior in stock and commodity prices. The 20-day average is popular. It is obtained by averaging each day the closing prices for the last 20 days, discarding the oldest price as each new price is added.

## AVERAGING PERCENT CHANGE

How do we compute the "average" value of sequential changes in percent gains and losses? First of all, how do we not do it? For sure, we do not use the arithmetic average. That should not surprise you. You know that if something you own loses 50% of its value, e.g., drops from $10 to $5, a subsequent gain of 50% won't get you even. It must increase 100% to return to its initial value. Let's find a system to define averages when percentages are involved.

Assume your mother says that the price of your home has increased by 26% over the past three years. We first convert the percentage into a multiplier. We say that today's value is 1.26 times its value three years ago. We ran into things like this in Chapter 7. Now, what was the "average" increase per year? In other words, what multiplier three years in a row would have produced 1.26 overall? Obviously, the cube root of 1.26, which is 1.08. If something increases 8%/yr for three years, its compounded increase is 26%. Do you know what has happened? We have stumbled onto the "geometric" mean.

The more common case is one in which you are given a series of percent changes and are asked to compute the compounded value. For example, the census in your school changed by +5% the first year, +2% the 2$^{nd}$ year, −3% the 3$^{rd}$ year and +4% the 4$^{th}$ year. What was the average annual change? My calculation shows:

$$[1.05 \times 1.02 \times 0.97 \times 1.04]^{1/4} = 1.0195 \quad \text{or } 2\%/\text{yr}$$

Putting this into a general formula:

$$\text{Geometric Mean} = [a_1 \times a_2 \times a_3 \times \ldots a_N]^{1/N}$$

Observe that the order of occurrence of the changes does not affect the average. This may seem surprising.

We use the geometric mean to average things that multiply together, just as we use the arithmetic mean to average things that add together. It's all a matter of using the right tool for each job.

# 11
# Pi ALL OVER

You are familiar with the Greek letter pi $\pi$ from its use as the symbol for the constant used in computing the area or circumference of a circle. What you may not know is how widely this unusual constant appears throughout mathematics and the applied sciences.

The 5th century B.C. Greek mathematician, Hippocrates of Chios (not the physician, Hippocrates of Cos), proved that the area of a circle is proportional to the square of its radius $r$. Although he did not define the constant of proportionality, he is credited with discovering that such a constant exists. Not unexpectedly, $\pi$ appears in any number of related geometric formulae. Use of $\pi$ as a symbol dates back only to about 1750. We always use '$\pi$' in formulae, never '$pi$' because it can be misinterpreted as $p \times i$.

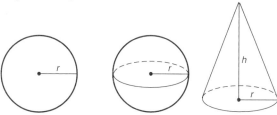

Circle Area = $\pi r^2$    Circle Circumference = $2\pi r$
Sphere Area = $4\pi r^2$    Sphere Volume = $4/3\, \pi r^3$
Cone Area = $\pi r h$    Cone Volume = $1/3\, \pi r^2 h$

What is the value of $\pi$? Well, in the 3rd century B.C., Archimedes estimated it to be somewhere between 3 1/7 and 3 10/71. The Chinese were much closer at 355/113. Check and see just how close. Punch up π on your calculator. Mine reads 3.141592654. Computers have carried it out to billions of places. It will never terminate or repeat endlessly. Here it is to a few more places than on your calculator:

$$\pi = 3.14159265358979323846264338 3279...$$

How can you or I determine the value of π? Using a little geometry, we can inscribe a polygon inside a circle and compute its perimeter from its component triangles. Since the legs of the triangles are identical to the radius of the circle, we can compute π for polygons with greater and greater numbers of sides, $N$. This is an excellent exercise (see pg. 54). By my derivation, the perimeter ($2rN \sin 180/N$) of the polygon approaches the circumference, ($2\pi r$) of the circle. And therefore $\pi \approx N \sin 180/N$. For $N = 2000$, $\pi = 3.1415913\ldots$ Test for other numbers of sides, just to get the feel for the convergence.

There are also any number of series that we can evaluate. They date back to the 16th century, 1900 years after Archimedes. The simpler ones include:

$$\pi/2 = (2 \times 2 \times 4 \times 4 \times 6 \times 6 \times 8 \ldots)/(1 \times 3 \times 3 \times 5 \times 5 \times 7 \times 7 \ldots)$$

$$\pi/4 = 1 - 1/3 + 1/5 - 1/7 + 1/9 - 1/11 \ldots$$

$$\pi^2/6 = 1/1^2 + 1/2^2 + 1/3^2 + 1/4^2 + 1/5^2 + \ldots$$

Try evaluating one of them. Those I tried converged so slowly that I had to accept prior proof that they converge to the value of π. How difficult it must have been to discover and prove them when you consider that the work predated modern computers.

Just think about it. Pi does not have a precise value, yet it can be expressed in terms of various series. It occupied mathematicians for more than 2000 years as they tried to unravel its secrets. Not until 1882 was π proved to be a transcendental number.

Without another assignment, pi might have been relegated to calculating areas and volumes. But its application to angular measurements made it ubiquitous throughout the sciences.

## THE RADIAN

One property of a circle is that when its radius $r$ rotates through an angle $\theta$ it describes an arc of length, $r\theta$.

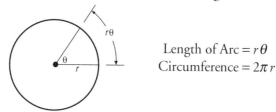

Length of Arc = $r\theta$
Circumference = $2\pi r$

For full rotation, the arc equals the circumference, in which case $\theta = 2\pi$. We have a name for this unit of rotational angle. We call it the radian. You are used to calling full rotation 360°. Well, in radians:

$$2\pi \text{ radians} = 360°$$
$$\pi \text{ radians} = 180°$$
$$\pi/2 \text{ radians} = 90°$$

In the applied sciences we have largely abandoned the degree, a special 1/60 counting system dating to the Sumerians of 5000 years ago, in favor of the radian.

You may also recognize the combination '$r,\theta$' as the basis for polar coordinates (an alternative to rectilinear coordinates) in which we identify a point by its distance and rotational angle with respect to a reference:

Now visualize an on-going sine wave. We can divide each complete cycle into successive increments of $2\pi$ radians, starting at time zero. This produces a situation in which $\pi$ is a measure of time. Does $\pi$ begin to remind you of the mighty oak that from a little acorn grew?

* * *

## ESTIMATION OF $\pi$

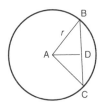

Triangle ABC is a component of inscribed polygon of $N$ sides.

Angle BAC = $360°/N$

Angle BAD = $180°/N$

$$\sin\frac{180°}{N} = \frac{BC/2}{r} = \frac{BC}{2r}$$

$BC = 2r\sin 180°/N$

Circumference ($2\pi r$) $\rightarrow N \times BC$, as $N$ becomes large

$\rightarrow N \times 2r\sin 180°/N$

$\therefore N\sin 180°/N \rightarrow \pi$, as $N$ becomes larger

# 12

# DESTINATION MOON

What more exotic venture could anyone devise than to escape from the confines of spaceship Earth and travel off into the vastness of space! Mankind has always had a spirit of adventure. Marco Polo faced unknown lands and unknown peoples during his 24-year adventure in the Orient. Christopher Columbus braved strange seas to seek a direct water route to the Far East. Now, in recent years, Americans overcame the uncertainties of rockets, computers and life support systems to travel to our nearest neighbor in the solar system.

One thing we have that early adventurers did not have is an accurate map. It is said that Columbus didn't know where he was going and when he got there, he didn't know where he was. But our route to the moon and back is precisely laid out.

To free our Apollo spacecraft from the grasp of Earth's gravitational field, we must accelerate to a speed greater that 25,000 MPH – called our "escape" velocity. Scientists could have accomplished this in one series of rocket firings. But it is much safer if we first go into Earth orbit, check that all systems are functioning properly and then exit into a trajectory to the moon. This "route" also requires less rocket power. Just as a discus thrower spins around and around to build up momentum before releasing a discus, so too will we take advantage of our orbital momentum around Earth to give our spacecraft an added boost toward the moon.

So let's be on our way. We are at Cape Canaveral, Florida. Our spacecraft is perched atop a giant Saturn 5 rocket, almost as tall as a 40-story building. We are strapped into our couches, three abreast and facing upward, as the final countdown continues. "T minus five seconds and counting, T minus four seconds, T minus three, T minus two, T minus one. IGNITION."

Low-frequency vibrations tell us that our first-stage engines have ignited. Our takeoff weight is 6.5 million pounds. The engines rapidly build up thrust, pushing harder and harder against the launch pad. When the thrust reaches 6.5 million pounds, one could theoretically lift the whole structure with the little finger. But we are still tethered, just as a propeller airplane often locks its wheels and revs its engines until ready to roll. At 7.5 million pounds of thrust we are released for LIFT-OFF.

Slowly we begin to rise – just three-quarters of an inch in the first half-second. Within ten seconds we clear the launch tower and two seconds later pitch forward into a graceful upward arc toward Earth orbit.

Amazing factors are at work. Most of our takeoff weight consisted of fuel which our engines are burning at the enormous rate of about two million pounds per minute. So, as the seconds pass, we are getting lighter and lighter, enabling the thrust of our lift-off stage to propel us ever faster and faster.

At the three-minute mark we are up 40 miles and traveling 6000 MPH. Five seconds later we jettison our liftoff rocket and ignite our second stage. Our weight is down to 1.5 million pounds, so our second-stage rockets are much smaller. Yet we double our speed in less than five minutes. Then we jettison this stage and ignite our third stage. We are GO FOR EARTH ORBIT.

Entering orbit requires delicate balance. If our velocity is too fast, we will drift off into space. If too slow, we will fall back to Earth. But when the outward force caused by our motion around Earth exactly balances the inward pull of Earth's gravity, our spacecraft keeps falling around Earth.

You can demonstrate these effects by tieing a rope to a pail of water. Swing the pail in an orbit perpendicular to the ground. Feel the force created by the weight. Swing it faster and the pull becomes stronger. If you let go, the pail will zoom away. Now let the speed slowly decrease. The force gets weaker and weaker until, finally, when the pull becomes smaller than the force of gravity, the pail falls out of orbit.

Now, back to our adventure. We orbit at an altitude of 100 miles, have a forward speed of 18,000 MPH, and circle Earth every 90 minutes. After being pinned to our couches by forces as great as 4.5 g, we are suddenly in a state of apparent weightlessness. There is no up or down. You have undoubtedly had this feeling on a roller coaster or in an automobile when you pass over a rise at such a speed that you are thrust upward by a force just equal to your weight. For a moment you have that eerie sensation of floating in space.

We are in a parking orbit – a safety stop where we can move about; adapt to our new state of weightlessness; and check out our guidance systems, environmental controls and communications with our controllers in Houston.

Houston maintains tracking stations in Spain, Australia and California, one of which will always have line-of-sight communication with us except for periods when we will be behind the moon. In addition, six instrumentation aircraft are aloft around the world as backup in case of ground station failure.

Half way through our second orbit, we go for our lunar trajectory. At a precise point in space and for a precise duration of 5 3/4 minutes, we refire our third-stage rocket and then jettison it. We rise farther above Earth, accelerate through our escape velocity of 25,000 MPH and are catapulted out of orbit. In another three minutes we are 1400 miles from Earth. At fourteen hours from liftoff we are 66,000 miles out and speeding through space at 48,000 MPH to rendezvous with the moon, 240,000 miles away, on our fourth day.

If you had been Frank Borman, William Anders or James Lovell aboard Apollo 8 on December 21, 1968 you would have experienced our first escape from Earth, on a shakedown cruise to orbit the moon. But we are aboard Apollo 11 on July 16, 1969 (see For All Mankind) and we will attempt the first landing on another "planet."

Our voyage is an indescribable experience. Space is unbelievably black. The stars don't twinkle, because there is no atmosphere. There is no way to sense speed, except as we look back toward Earth, we see less and less detail. Eventually we see our Blue Planet in its entirety and then it appears smaller and smaller. We see it as it truly exists in space. We feel a cosmic separation from Earth. Why shouldn't we? We are extraterrestrial beings! A UFO, perhaps, to the Man in the Moon?

We maintain a regimen to keep our bodies functioning in 24-hour days to which we are accustomed, for in space there is no natural day/night cycle. Our work and sleep cycles are based on Houston time. Houston awakens us each "morning" with news headlines and updates on our flight plans.

At the 24-hour mark we are 100,000 miles from Earth – almost half way to the moon. A complex system of celestial gravitational forces has reduced our speed to 3600 MPH. There has been sufficient drift from our course that we carefully fire our thrusters for a three-second, mid-course correction.

We are getting closer to the moon, but we cannot see it because our trajectory has it aligned so closely with the Sun that we are blinded by solar glare.

By late afternoon of Day 3 we pass the 186,000 mile mark, with our speed reduced to 2000 MPH. By now we are experiencing noticeable delays in our communications with Houston. Since radio transmissions travel at the speed of 186,000 miles per second, the transit time is now greater than one second in each direction. That can be a long time when awaiting critical information or instructions. And the delay becomes longer as we get further from Earth.

That night, at 63 hours into our mission, we enter the lunar sphere of gravitational attraction. Virtually imperceptibly we begin to accelerate again. We are still 40,000 miles from our intended landing site.

The morning of Day 4 finds us preparing for insertion into lunar orbit. We will have only one try at slowing our spacecraft enough to be gravitationally curved into orbit. This burn is especially risky because it will take place during the loss-of-signal period while we round the back side of the moon. We are on our own to make the burn. Houston cannot assist us. If we do not slow down enough, we will sweep past the moon and on toward outer space. Computers give us the countdown for firing, but one of us must throw the switch. The world will not know the outcome until we successfully emerge from the other side of the moon. Finally, we are no longer blinded by the Sun. We get our first good look at the moon, its surface brightly illuminated by "Earthshine."

Our first orbit is not circular, so we perform another brief burn to circularize it at 60 miles altitude with a forward speed of 3600 MPH. It takes two hours to orbit the moon, of which 34 minutes is on the back side out of contact with Houston. As busy as we are, we still have time to enjoy the immense satisfaction of our accomplishments thus far and, when behind the moon, the solitude of being totally out of contact with everyone else in the universe.

It is Day 5. We undock our Lunar Excursion Module (LEM) and prepare, after thirteen orbits, for descent to the surface. So LEM, named Eagle, with Neil Armstrong and Edwin "Buzz" Aldrin aboard separates from spaceship Columbia with Michael Collins aboard. Eagle is very small, indeed, Neil and Buzz having to stand as they pilot it to a soft landing on the Sea of Tranquility where Neil announced to all the world 'THE EAGLE HAS LANDED." July 20, 1969. Several hours later when he became the first human to step on the moon, Neil made his epic statement, "THAT'S ONE SMALL STEP FOR MAN, ONE GIANT LEAP FOR MANKIND."

Television gave this historic event broad coverage around the world. An estimated one-sixth of Earth's population was watching. I was vacationing in Greece at the time. Recognizing us as Americans, a young boy came up and exclaimed that we had landed on the moon. We asked how he knew. "On TVs in front of the American embassy," he replied.

They planted a plaque, made geological observations and took rock samples, during a 2 1/2 hour walk on the moon. Others would follow. By the time the Apollo program ended in 1972, there were six landings and explorations in diverse locations, ranging from a mountainous area to one of ancient volcanic activity. While twelve brave astronauts had the historic opportunity to land on the moon, six others kept lonely vigils in their motherships.

Now it is time to lift off. LEM has only one ascent engine, the same one that braked us for our landing. It is said to have been the only key component that lacked a backup. It had to ignite on time and burn long enough to boost us into lunar orbit, or else! We rise to a 60,000 ft. parking orbit from which we later ascend to rendezvous with Columbia.

The Apollo 11 astronauts. From left to right, Neil Armstrong, Michael Collins and Edwin (Buzz) Aldrin, Jr.

You can imagine the latent concern that even these experienced test pilots must have had about how perfectly the hardware had to perform at so many critical junctures of their voyages. There were numerous serious malfunctions and one near-disaster, but all returned safely. Sadly, three men including the well liked Gus Grissom died in a test stand exercise in 1967. They were the only casualties. A superb record for such an ambitious program.

We rendezvous and dock after a precarious game of tag that takes several hours, with Michael Collins in Columbia executing the intricate maneuvers all by himself. All that is left to do now is to jettison Eagle and blast Columbia out of lunar orbit and on its way back to Earth – again, doing this from behind the moon. Because of the moon's lower gravity, our escape velocity is only 5300 MPH. The three-day trip home seems long because there is so little activity. We have ample time to ponder our profound experiences. Finally, we prepare for our final burn and splashdown in the Pacific Ocean.

Imagine being one of those astronaut heros. I'll bet you would gaze at the moon each night of your life and proudly say, "I was there." You would remember the events as if they happened only yesterday, yet mixed with these emotions would probably be a latent feeling of disbelief.

Many years have passed since these historic missions. Since then we have sent unmanned spacecraft to distant Jupiter, Neptune, Uranus and beyond to who knows where. But getting astronauts to nearby Mars, a possible project in the 21st century, will make today's accomplishments seem easy.

# 13
# RADIATION ALL AROUND

We are bathed in an extraordinary range of electromagnetic radiation. Much of it comes from outer space, especially from the sun. The rest we generate here on Earth. Count the number of decades on the frequency scale in the graphic on the next page. There are 25. This means a range of $10^{25}$ from one end of the radiation spectrum to the other. Review Chapter 4 to see what a huge range this is. Guess what? We are exposed to energy throughout much of this range every day of our lives. A rudimentary knowledge of the radiation process will help you understand both radiation itself and many scientific developments you read about in the news.

Until the time of Albert Einstein in the 20th century, scientists believed that outer space had to be filled with a transmitting medium (which they called "ether"). This gave them a propagation model similar to the one which carries our voices through the air (sound being unable to travel in a vacuum). They eventually learned, however, that electromagnetic energy does not require a transmitting medium. It travels through the vacuum of space and it does so at the speed of light. They also discovered that it does not flow continuously, but in short bursts of sinusoidal waves, called quanta or photons. This was a monumental discovery.

## HOW RADIATIONS DIFFER

The electromagnetic spectrum is scaled in both frequency and wavelength. Do you see that they vary inversely with each other? The higher the frequency, the shorter the wavelength,

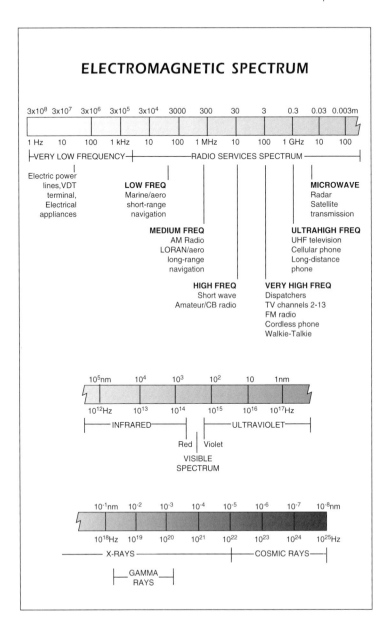

and vice versa. Also note that increments of 10-times are equally spaced. This indicates a logarithmic scale, doesn't it? As with other things we have studied, the underlying physical relationship is amazingly simple:

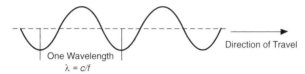

where $\lambda$ is its wavelength, $c$ is the speed of light (186,282 miles per second in the vacuum of space) and $f$ is the frequency of the wave.

Engineers identify radiation by its frequency, as on your radio dial or on a microwave oven's precautionary medical label. On the other hand, physicists identify light by its wavelength, as do amateur radio operators when they refer to the "100-meter band." So we present both scales.

Before we go on, let's note two other basic facts. The first is that velocity decreases when radiation leaves the vacuum of space and travels through our world of gases, liquids and solids. So, you might ask, how does this affect frequency and wavelength – one or both must change when velocity decreases. That's a very insightful question. The answer is that wavelength decreases and frequency remains unchanged. The second fact is that the higher the frequency, the greater the energy. Other than these differences, all electromagnetic radiation is identical.

## VISIBLE LIGHT

Our most important radiation is visible light, located in a very narrow band of the spectrum between about 750 and 390 nm. Visual perception peaks at about 550 nm. We see different wavelengths as different colors: 680 nm is red, 560 nm is yellow, 500 nm is green, 420 nm is blue, 400 nm is violet, plus all the colors in between. When light strikes an object, some wavelengths are absorbed and others are reflected. We see only the reflected energy. Grass is green, for example, because it reflects energy in the vicinity of 500 nm. And when apples ripen, they change their reflection characteristics, don't they?

If all visible energy is reflected, we see the object as being white. If all is absorbed, we see it as black. Since absorbed energy is converted into heat, the difference in absorption explains why a black car sitting in the sun is hotter than a white car and why we wear dark clothing in the winter and white in summer. Earth's atmosphere absorbs much of the sun's energy during the daytime and reradiates it at night. This is one reason our planet is so uniquely habitable. Planets with no atmosphere are intensely cold at night.

Our sun emits radiation at almost all wavelengths, about half of it in the narrow visible band. It peaks in the vicinity of yellow light. Isn't it interesting that our eyes and those of most animals are tuned to this narrow segment of the spectrum? Is this a coincidence or have we adapted over time?

The sun's yellow color is characteristic of its temperature, which is about 10,800° F. Hotter stars tend toward violet; cooler ones toward red. Similarly, skilled workers in steel mills judge the temperature of hot ingots by their color.

## INFRARED

From the boundary of red light to about 1 mm ($10^6$ nm), is the infrared (IR) range. This spectral region is very interesting. Our bodies emit radiation in this range. So do exhaust gases and machinery, as well as plants and animals. Frequency is lower because the temperatures are lower.

We cannot see IR radiation directly, but scientists have designed goggles with built-in IR sensors that enable soldiers to "see" at night by presenting patterns of objects that are hotter than the background. Many lives were saved in Operation Desert Storm because we could see the enemy in the dark of night, but the enemy could not see us. In a similar fashion, missiles employ IR sensors to home-in on the exhaust of enemy aircraft.

We use low-cost IR monitors to detect intruders as well as hotspots through which heat escapes from buildings. And IR sensors in orbiting satellites enable scientists to predict the size of harvests and to detect illegal crops.

Finally, the IR region is very efficient for heating purposes. At room temperature, molecules in our bodies and other objects vibrate at the same frequencies as IR energy. This makes it easy for these molecules to absorb energy from an infrared lamp, making them vibrate more rapidly and create heat through friction. Heating is accomplished independently of the surrounding air temperature. We use infrared heaters to keep food warm and to make us comfortable while dining outdoors in cool weather.

## ULTRAVIOLET/X-RAYS/GAMMA RAYS/COSMIC RAYS

Shorter wavelengths than visible light, from the boundary of violet light to about 1 nm, comprise the ultraviolet (UV) region. As you know, it is the sun's UV rays that can burn our skin and damage our eyes. Here we begin to see the harmful effects of the higher energy possessed by high-frequency radiation. In the computer field we use UV light to etch circuits on computer chips. Newer machines employ X-rays to make even finer lines for greater circuit density – the shorter the wavelength, the finer the line.

In the UV range, it becomes difficult to attribute radiation to a body's heat because few stars are hot enough to produce such intense energy. We need new explanations for the source of energy.

The Danish physicist, Niels Bohr, gave us one answer in

1913 when he discovered that a quantum of energy is released whenever an electron changes its energy state by jumping from an atom's outer orbit to an inner orbit. Simultaneous emissions from many billions of atoms cause the characteristic radiation wavelengths of chemical elements. Helium was first identified by a spectrum line during an eclipse of the sun in 1868. It showed up as a wavelength never observed before. Then scientists set out to find it in the physical world.

We cause chemical elements to radiate in this manner, as when we stimulate emission from mercury for use in sunlamps. It also occurs in the visible range, as in sodium street lights, fluorescent lamps and all those light-emitting diodes (LEDs) found on home appliances.

In the X-ray region, photons of energy are produced by rapidly decelerating a stream of electrons. The greater the rate of deceleration, the higher the frequency and energy. We fire X-rays at the human body to probe for tooth decay, bone fractures and tumors.

Gamma Γ rays are produced in the nucleus of atoms when protons change their energy state. They are released by nuclear reactions. Finally, we have cosmic rays which are charged particles, especially protons and electrons, from outer space. Few reach Earth's surface. However, as they pass through the upper atmosphere, they collide with atmospheric nuclei and blast them apart, creating a shower of secondary cosmic rays.

## MICROWAVES/RADAR/TV/RADIO/LORAN

Now let's go to longer and longer wavelengths, or lower and lower frequencies. We will be considering electromagnetic energy created here on Earth. Discussion will be easier if we start with the low frequencies and work our way upward.

Let's start with 60 Hz radiation from electrical appliances, power lines, and video terminals. Check their wavelength. It is on the order of 3000 miles. Imagine that! Very low energy, isn't it? Concern has arisen over potential health hazards. Anything is possible, but we cannot attribute it to the energy of radiation. If it does affect human cells, a yet unknown mechanism is at work.

Low-frequency, land-based LORAN transmitters enable ships to continuously identify their latitude and longitude to within a matter of feet. The cost of electronics is now so economical that in coastal waters we find LORAN receivers on most pleasure boats. They are now being replaced with Global Positioning Systems.

To avoid aircraft collisions and to assure safe landings in bad weather, we transmit radar waves and track their reflections from aircraft. For smaller objects we need shorter wavelengths, so we use microwaves to track small missiles. We also use microwaves to heat food by exciting molecules in the way described for infrared heat.

We also do something not previously discussed, and that is to use radiation as a "carrier" to transmit lower-frequency information. We use AM and FM radio frequencies to transmit music and voice. Television frequencies transmit visual images. From here we advance to microwaves to transmit large numbers of TV and telephone channels.

## OTHER FORMS OF RADIATION

In addition to electromagnetic radiation, we have acoustical fields of underwater acoustics, sound and ultrasonics. We use sonar to measure water depth and to locate schools of fish. Loudspeakers deliver radio and TV sounds. Ultrasonic systems enable doctors to identify the sex of unborn babies, and they provide energy for cleaning objects ranging from jewelry to teeth to delicate instruments.

Here we use the mechanical motion of loudspeakers and ceramic elements to impart vibratory motion to water, air and other transmitting media. The same wave formula applies, except that velocity is very much slower than the speed of light. The speed of sound is about 0.95 miles/second in the human body, about 0.92 miles/second in sea water and about 0.21 miles/second in air. These wavelengths are very much shorter than electromagnetic waves at the same frequencies, aren't they?

Students specializing in the study of radiation are assured a lifetime of challenging and gratifying work involving health care, communications and public safety. Once students understand wave propagation, they can work in any of these fields.

# 14
# WHERE'S THE ETHER!

Until a century ago scientists accepted the age-old theory that every form of energy requires a medium in which to propagate. Knowing this to be true for sound, vibration and earthquakes, the ancients naturally assumed it to be true for light. They gave this substance the name "ether" or "aether." The idea began with Aristotle when he postulated the existence of ether in the regions above the lunar sphere. By the 14th century it was assumed to be everywhere in the cosmos.

Scientists developed a list of the physical properties that ether must have. 1) It must have no mass. Good thing, because if it had the slightest mass it might account for the dark matter so important to determining the expansion characteristics of our Universe. 2) It must offer no resistance to the motion of objects through it. Otherwise planets would gradually lose speed and change orbit, even drop out of orbit. 3) It must be very stiff. Because light travels so fast and because the stiffer the medium the faster that waves can travel through it, the assumed stiffness had to be huge. It was a wondrous substance indeed! This alone makes one wonder why these brilliant men did not seek an alternative explanation for how light propagates.

The notion of the existence of ether required an important conclusion. It was that if light propagates through a medium as does sound through air, its speed must be a function of the relative speeds of the medium and the receiving body. Liken it to a motorboat running on a river at constant throttle setting. Using the riverbank as a frame of reference, the boat will move faster downstream than upstream.

## MECHANISM OF ELECTROMAGNETICS

By the 1850s, scientists were struggling with a nagging fundamental question as to the mechanism by which one electromagnetic charge can react across space and exert a force upon another charge. Magnets and transformers are good examples. They called it the "action-at-a-distance" problem. Interestingly, we are experiencing the same struggle today with gravity. We do not know the mechanism by which gravity works.

The leading character was James Maxwell (1831-1879), a Scotsman, who discovered the laws of electricity and magnetism. He was also interested in light. He determined that light represents just a very small portion of the vast electromagnetic spectrum (Chapter 13). He also showed that the electric ether, magnetic ether and luminiferous ether should all be one and the same. Twenty years later Heinrich Hertz added radio waves to the list.

Maxwell measured the speed of light and also calculated its speed, using a formula that did not include the properties of a medium. His equations showed that the speed of light is constant no matter how a body moves in space. This contradicted, but did not rule out, the existence of a medium. So work continued with a long series of unsuccessful telescopic experiments to detect the ether. The stage was set for someone with new experimental ideas. Ironically, the goal was still to prove rather than disprove the existence of the ether.

## MICHELSON-MORLEY EXPERIMENT

On the scene in 1881 came a twenty-eight-year-old American named Albert Michelson. He knew that Earth moved in its orbit around the sun, and therefore through the assumed ether, at a speed of about 18 miles per second. He also knew

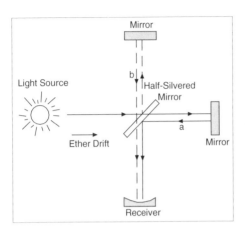

that he could treat this as a reverse situation in which Earth is at rest and the ether moves past it.

Michelson's goal was to compare the speed of light in the direction of Earth's motion through the ether with its speed perpendicular to Earth's motion. He invented an optical interferometer technique in which an L-shaped apparatus splits a beam of light into two beams and guides them along separate perpendicular paths of identical length – one beam (a) down and back along the path of the incoming light train and the other (b) perpendicular to it – and then recombines them at the receiver. Taking into account the hypothetical ether-drift velocity on the speed of light, trigonometry shows that the time required to traverse the downstream-upstream path will be slightly longer than that of the cross-stream path. This results in the two beams being out of phase when recombined, thereby forming interference fringes (bands of alternately light and dark color) from which it is possible to calculate the speed of the ether wind in relation to the movement of Earth. These comparisons required his equipment to resolve time differences as small as five parts in a billion. To his great surprise, he found no difference between the round trip times of the two wave trains. We call this a "null effect." He considered the experiment a failure because it did not support the theory he had set out to prove – that ether existed or could exist. And, as with Maxwell's work, it did not prove that ether could not exist.

Michelson, now a professor of physics at Case School of Applied Science, was undaunted. He teamed up with fellow scientist Edward Morley, noted professor of chemistry at nearby Western Reserve University. They did everything they could to make their experiment more conclusive. They protected against vibration by mounting the interferometer on a stone slab and floating it on a film of mercury, achieved 40-times the required measurement resolution, and performed the experiment at different times of the day and year. But in 1887 they still experienced the null effect! The time had come to run it up the flagpole and see if anyone saluted it! This result presented the world's greatest scientists with a profound dilemma. They were forced to believe one of the following: 1) Earth is at rest in space – an unthinkable assumption at that late date. 2) Ether is dragged along by Earth and causes the two to have no relative motion – already ruled out by a number of experiments. 3) Ether does not exist. To many of them this was the equivalent of scrap-

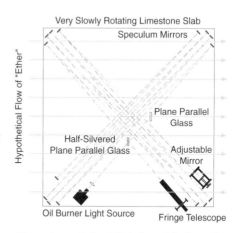

*Plan view of the Michelson-Morley ether drift experiment. The apparatus was mounted on a 5-foot square, 1-foot thick limestone slab which was supported on a doughnut-shaped bearing lubricated with mercury. After adjustment, interference fringes of the perpendicular light paths could be observed at the telescope. The apparatus was set into very slow rotation. If the ether flow had any effect on the light paths, the observed interference fringes would change. "Unfortunatley," they did not. Thus the ether was considered nonexistent.*

ping their views of electromagnetism. 4) They were missing an unknown feature of the physical world.

The failure of the Michelson-Morley experiment to prove the existence of ether is one of the most successful "failures" in the annals of science. It proved that light exhibits a uniform speed regardless of its direction or the speed of an observer. In other words, the speed of light is constant. It would correlate well with the theory of relativity that came shortly thereafter.

*Albert A. Michelson (left) and Albert Einstein at the California Institute of Technology, January 1931.*

On Friday, June 30, 1905 Albert Einstein submitted a 31-page paper titled, "On the Electrodynamics of Moving Bodies" to the journal, *Annalen der Physik*. It was published three months later. It implied that ether could not be detected experimentally and was therefore a useless concept that should be discarded. Einstein later wrote, "This ether had to lead a ghastly existence alongside the rest of matter, in as much as it seemed to offer no resistance whatsoever to the motion of ponderable bodies." There was no longer any reason to argue for the existence of that magnificent substance.

## Albert Michelson

Albert Abraham Michelson (1852-1931) was born in an area of Germany that is now part of Poland. His family emigrated to the United States when he was two years of age and settled in Nevada. He graduated from the U.S. Naval Academy with outstanding grades in science and soon returned as a permanent lecturer in physics. He attended universities in Berlin, Heidelberg and Paris to study optics, which he perceived as essential to his ambition of measuring the precise speed of light.

He became a professor of physics at Case School of Applied Science in Cleveland, Ohio, where in 1882 he resumed his efforts in measuring the speed of light. He announced that year the most accurate speed to date, one that stood for 30 years until he revised it himself. In another optical experiment, he and Professor Morley (1838-1923) used the red spectrum of cadmium to redefine the length of the standard meter in Paris. On December 10, 1907 he became the first American to win the Nobel Prize in science for: "contributions in interferometry and optics and for precise measurements of the speed of light." He went on to become professor of physics at Clark University 1889-1892 and head of the department of physics at the University of Chicago (1892-1929).

Albert Einstein paid public tribute to Professor Michelson in 1931 for his extensive contributions to science: "My honored Dr. Michelson, it was you who led physicists into new paths and through your marvelous experimental work paved the way for the development of the theory of relativity."

# 15
# MENDELEEV'S NUMBERS

If you are fascinated by the manner in which numbers enable us to find order in the solar system, or by the way that numbers help us describe radiation, then the discoveries of Mendeleev and his successors will amaze you even more.

Dmitri Ivanovich Mendeleev was born in Siberia in 1834. He studied in St. Petersburg as well as in France and Germany before earning his doctorate in chemistry. After years of studying and experimenting with chemical elements, he became obsessed with the idea of finding a systematic method of cataloging them. Along the way when he took time off to help his country solve its educational problems he wrote, "We could live at the present time without a Plato, but a double number of Newtons is required to discover the secrets of nature and to bring life into harmony with its laws." What a marvelous lead-in for the theme of this chapter!

From time immemorial, artisans have fashioned jewelry, tools and weapons using carbon, copper, lead, mercury, gold, silver, iron and tin. Whether these were recognized as elements is another question. After all, Aristotle believed that all matter was composed of earth, air, fire and water. By the time of the Renaissance, eager chemists throughout the world were discovering new elements – antimony and bismuth in Germany, platinum in Colombia, cobalt and magnesium in France, nickel and manganese in Scandinavia. But how many elements were there?

Around 1800, John Dalton in England observed that all matter is composed entirely of a relatively small number of chemical elements. By the time of Mendeleev, some sixty-three had been found. The only thing they seemed to have in common was their diversity. Some are gases; a few are liquids; most are solids – at room temperature and pressure of course. Phosphorus, potassium and sodium are so volatile they burn your skin and they cause a real spectacle when they react with water; most are much less active. Gold never tarnishes; iron rusts very quickly. Copper, gold and silver are very malleable; platinum and iridium are extremely hard. Nickel and platinum take a high polish; lead and aluminum are dull. What a hodgepodge of data! But each element did have one unique property – its "atomic weight." Might this lead to the discovery of some order among the elements?

Unknown to Mendeleev, John Newlands in England had observed that when the known elements were arranged in order of their weight, about every seventh one had similar properties. When in 1865 he gave a paper likening this periodicity to the octaves on a piano keyboard, members of the English Chemical Society laughed at him. "Too fantastic," they argued. One member jeeringly asked if Newlands had ever examined the elements according to their alphabetical order, since any arrangement can present some random coincidences.

Mendeleev observed the same phenomenon, but he had a more imaginative mind. He searched for other things that might be going on. He made sixty-three cards and wrote on each the name and properties of one of the known elements. This was not an easy task. He corresponded with hundreds of chemists in many countries to accumulate his data and he had to perform additional experiments himself. He even included a card for fluorine whose presence was recognized but had never been isolated because it is so volatile.

Like a child at play, Mendeleev pinned the cards to his laboratory wall and studied them over and over again, sorting out similar elements and grouping them in different ways, relentless in his effort to find some order. He hit upon the idea of placing the elements in a row across his wall, in the order of their relative atomic weights. But first he had to deal with hydrogen, the lightest element. Since it had no common properties with the rest, he set it aside at the top.

Below it he placed lithium followed in a row by beryllium, boron, carbon, nitrogen, oxygen and fluorine. The next element in atomic weight was sodium (Na). It had properties similar to lithium, so he placed it below lithium and started a new row that continued with magnesium, aluminum, silicon, phosphorus, sulfur and chlorine. Each had properties similar to the element above it. Next came potassium (K). Did it begin a new period by displaying properties similar to lithium and sodium? It did, so he began a third row. He placed calcium in the next position because it had similar properties to magnesium and beryllium.

| $H_1$ | | | | | | | |
|---|---|---|---|---|---|---|---|
| $Li_7$ | $Be_9$ | $B_{11}$ | $C_{12}$ | $N_{14}$ | $O_{16}$ | $F_{19}$ | |
| $Na_{23}$ | $Mg_{24}$ | $Al_{27}$ | $Si_{28}$ | $P_{31}$ | $S_{32}$ | $Cl_{35}$ | |
| $K_{39}$ | $Ca_{40}$ | | $Ti_{48}$ | | | | |

*Mendeleev's elements and their relative atomic weights*

Mendeleev must have experienced great satisfaction as he studied his "periodic" table. Excluding hydrogen, what he saw developing were seven columns, each representing a family of elements that exhibited similar properties. Was this the order he was seeking? We shall see. His next element was titanium (Ti). It did not fit the boron and aluminum family, but it was similar to silicon and carbon. What did this suggest? A missing element yet to be discovered? Correct.

At this point I am reminded of an anecdote about a professor. When asked how he engaged his audience he replied,

Dmitri Ivanovich Mendeleev (1834-1907) in the chemical laboratory, University of St. Petersburg.

"First I tell them something they know, and that gains their confidence. Then I tell them something they don't know, and that gets their attention. Then I tell them something they don't believe, and that leads to a lively discussion."

Well, in 1869 when Mendeleev delivered his paper, "On the Relation of the Properties to Atomic Weights of the Elements," he gave members of the Russian Chemical Society plenty of things to disbelieve! He had finished arranging the known elements as best he could in his Periodic Table of the Elements. He told them that many elements were yet to be discovered. He even had the audacity to identify three of them, predicting their atomic weight and other properties. One was scandium, which follows calcium. The others were gallium and germanium. If that was not enough, he also stated that the atomic weights currently assigned to gold and tellurium were incorrect. This was downright heresy! But without these corrections, the elements did not fit the order of his table. Truth was on his side, and Mendeleev's position was assured in history.

By the year after his death in 1907, eighty-six of the ninety-four naturally occurring elements had been discovered. The rare gases were found and they formed an eighth column. And below the second row, a subset of smaller families of "transitional" elements were discovered between the metals of the first two columns and the nonmetals of the last six columns. Ironically, in his lifetime Mendeleev was never as famous in Russia as he was in the rest of the world. The tsar had punished him for his liberal politics by preventing his election to the Russian Academy of Science.

Dmitri Mendeleev was indeed a genius and he did wonders for the advancement of the science of chemistry. But in every age – even today – scientists of genius are limited by their basic, fundamental knowledge. Fortunately for all of us, the search for new knowledge never ends. Every answer to a question seems to raise a new question.

## ATOMIC NUMBERS

Mendeleev's work was complicated by the fact that there are 94 elements ranging in atomic weight from 1 to 244, and there is anything but an equal increment in weight from one to the next, as you can observe in his table. The operative word here is "atomic." It dates back to 400 B.C. when the Greek philosopher Democritus suggested that the world was made of a few basic building blocks too small for the eye to see, called atoms. The atom was still considered to be the fundamental, indivisible part of a chemical element. The time would come when someone would disprove this age-old assumption.

The critical breakthrough came in 1897 at Cambridge University in England when J.J. Thomson discovered the electron and in so doing proved that the atom is divisible. Key to our discussion is the fact that hydrogen has one electron, lithium has two electrons, etc., up to plutonium with 94 electrons, with no irregularities in the series. We call this structural characteristic of each element its "atomic number." It – not atomic weight – is the real basis for the order of the periodic table.

How amazing! These elements (see Periodic Table in most dictionaries) are the building blocks of all matter in the universe and the fundamental difference among them is simply an increase of one electron in their atoms as we progress from each element to the next. How much easier Mendeleev's work would have been, had he had atomic numbers with which to work.

As important as was Thomson's discovery of the electron, his former student, Ernest Rutherford, gave us the intellectual breakthrough that changed the world. In 1911, he proposed a new model for the atom – one in which electrons orbit a dense core (nucleus) in the manner that planets orbit the sun (but held together by electrical rather than gravitational forces.) In less that 300 years, the "outrages" of Copernicus and Galileo had become the model for modern chemistry and physics.

The electron is a tiny part of an atom's mass and it carries a negative charge. Physicists eventually discovered that successive families in the periodic table each had one additional electron in their outer orbit and this contributed to each family having similar properties. As to the transitional elements, they fall in the second family. The work that Mendeleev began was reaching completion. But what an incredible fact of nature! All matter in the universe is made up of one or more of the 94 elements, each characterized by a unique number of electrons, ranging from 1 to 94, orbiting its nucleus. And all of the elements exist on our planet! Few systemizations in science rival Mendeleev's periodic concept as a broad revelation of the order of the physical world.

We should not conclude this part of our story without mentioning elements heavier than uranium. Only minute quantities of neptunium (93) and plutonium (94) have been found in nature. Heavier elements only occur when synthesized in a laboratory, all elements above fermium (100) are very short-lived.

Physicists initially thought the atom was made up solely of electrons and an identical number of positively charged particles, called protons, inside the nucleus. But in 1932, James Chadwick found that the nucleus also contained other particles, called neutrons, which have almost the same mass as protons but have no electrical charge. This discovery changed atomic weight from a property they could previously only measure, to one that they could explain.*

An atom of hydrogen, the lightest element, contains one electron and one proton. It is unique; the rest of the elements also have neutrons. Since protons and neutrons (with a mass of approximately $1.67 \times 10^{-24}$ gram) are 1837- and 1838-times heavier respectively than electrons, they govern an atom's atomic weight.

Atomic number and atomic weight now become more meaningful to us. Let's pick gold (Au) and silver (Ag). Silver is the lighter of the two. You don't have to study chemistry to know that, do you? Its atomic number is 47, meaning that its atom has 47 electrons and 47 protons, correct? Its atomic weight is 108, meaning that it also has 61 neutrons. Correct? The corresponding numbers for gold are 79 and 197, meaning there are 79 protons and 118 neutrons in its nucleus. Now you know why gold is almost twice as heavy as silver.

Here is something else very interesting. In the periodic table, the next element above gold is mercury (Hg) with an atomic number of 80 and atomic weight of 201. If we could remove one proton and four neutrons from the nucleus of mercury, we would have gold! The ancient alchemists wanted to make gold from mercury. Maybe they weren't as crazy as we think.

So far, we have been talking only about the normal state of atoms. In some cases there are more or fewer neutrons than an atomic weight suggests. We give the name "isotopes" to these atoms. In general, different isotopes of an element behave in a similar manner chemically, but have slightly different physical

---

*Considering that atoms are not always electrically neutral, due to the possibility of some electrons being displaced from their orbits, it is more precise to identify the number of protons (rather than electrons) as the basis for the atomic numbers.

properties. Some are unstable, this condition being known as "radioactivity."

Let's put some more numbers on things. Consider hydrogen. About 99.98% of all naturally occurring hydrogen atoms have one proton and no neutrons. A few have one neutron. Called deuterium, or heavy hydrogen, it makes up about 0.0156% of all hydrogen atoms. It is used in some types of nuclear reactors to moderate the production of neutrons. Tritium, with two neutrons, represents about $10^{-15\%}$ of all hydrogen atoms. It is radioactive with a half-life of 12.26 years. Both isotopes are involved today in cold fusion experiments.

## THE FIELD OF CHEMISTRY

To say that chemistry is a vast field is a gross understatement. Everything in the universe is composed of chemical elements and the complex molecules that result from their unions, right down to the nucleic acids (DNA and RNA) by which information about the structure and function of living organisms is stored and passed on to successive generations.

Industrial advancement is heavily dependent upon chemistry: Lightweight, high-strength ceramic structures that make airplanes lighter and more fuel efficient. Ceramics for computer chips. Plastics. Long-life batteries. Pesticides and synthetic fertilizers. Food preservatives. Catalytic converters to reduce air pollution. And how about pharmaceuticals? What a field for research! Although more than 25% of all prescription drugs have their origins in plants, fewer than 1% of the world's flowering plant species have been thoroughly analyzed for potential medicinal value. Every industry depends upon chemistry, either directly or indirectly.

Perhaps the biggest potential of all is in biochemistry. There is undoubtedly more we don't know than do know about the human body. For example, the DNA has revealed fewer than one percent of its secrets. Most of medicine, even today, involves solving problems through surgery or the lifelong use of pharmaceuticals. Life-threatening diseases are seldom cured in the true sense of the word. In future years we may learn how to make the human body permanently reset its blood pressure, regulate its production of sugar, stop its production of bad cells, etc. Progress is beginning to be made through gene therapy. All we need are a few big breakthroughs. They may occur within your lifetime. Manage your finances well! You might live to be 110!

\* \* \*

## A SEQUEL TO ATOMIC STRUCTURE

We cannot leave the subject of atomic structure without updating our knowledge. Work in the early 20$^{th}$ century that gave us the electron, proton and neutron was yet again thought to have identified all the indivisible particles of matter. This belief was held until the late 1960s when physicists at the California Institute of Technology discovered that protons and neutrons are in fact made up of smaller particles. They named them "quarks." Quarks are emitted when protons or neutrons collide with other particles at high speeds. That's the kind of investigation done in particle accelerators – so-called atom smashers.

A proton is composed of two "up" quarks and one "down" quark. A neutron contains one up quark and two down quarks. The two kinds of quarks have almost the same mass, so this accounts for protons and neutrons also having nearly the same mass. But, you ask, with three quarks each, how can a proton have a charge of +1, and the neutron a charge of 0? Physicists have an answer. An up quark has a charge of +2/3, and a down quark has a charge of -1/3. So it all works out. To date, more than a hundred subatomic particles have been discovered.

Now, what about the electron. Is it an elementary particle?

They say it is. Are we finally getting to the bottom of this case? Maybe so. The renowned Stephen Hawking says there are theoretical reasons for believing that we have close to full knowledge of the ultimate building blocks of nature. But that's far from saying we are there. Even if we are, there is much work yet to be done in the related field of force-carrying particles. We know much less about them than we do the above matter-carrying particles.

## FORCE-CARRYING PARTICLES

Scientific theory currently groups force particles into four categories. The most well-known is gravitational force. Each of us can feel its effects, and it is the first force physicists detected, yet they know least about it. And it is the only force we cannot control. Not yet, at least. It is so weak that we would not notice it were it not for two important properties – it acts over long distances and it is always attractive. Consequently the very weak gravitational forces among all the particles of matter in two large bodies, such as Earth and the Sun, add up to immense forces.

We have never found the particle that carries gravitational force, but that has not kept us from giving it a name – the gravitron. Efforts to detect the gravitron are ongoing. Only after we thoroughly understand gravity, can we attempt to control it. Imagine how aircraft, helicopters, elevators and other equipment would change if we had antigravity equipment. Does this sound strange? Well, this is one way to account for the spontaneous motions we readily credit to UFOs!

The next category is electromagnetic force, which enables electrically charged particles like electrons and protons to interact. It is carried by photons. The electromagnetic force between two electrons is about $10^{34}$ times greater than gravitational force. Here we go again with the immense ratios present in nature.

On the micro scale of atoms, the electromagnetic attraction between negatively charged electrons and positively charged protons causes electrons to orbit the nucleus of an atom, just as gravitational attraction causes Earth to orbit the sun. But electromagnetic force is repulsive between like charges. On a macro scale, such as Earth and the sun, the attractive and repulsive forces between nearly equal numbers of positively and negatively charged particles cancel each other and there is little net electromagnetic force between the two bodies.

The final two categories are effective only within atoms. The strong nuclear force holds quarks together to form protons and neutrons and holds protons and neutrons together to form nuclei. We need it to overcome the repelling forces between like charges. It is carried by gluons. The more obscure weak nuclear force is responsible for certain forms of radioactivity and for the decay of heavy quarks into lighter ones. It is carried by bosons.

As if the exhaustive investigations of force-carrying particles is not enough, physicists the world over are working toward finding a grand unification theory that will explain electromagnetic forces, the weak and strong nuclear forces, and ultimately gravitational force as different components of a single force.

| Forces of Nature | | |
|---|---|---|
| | Range | Relative Strength |
| Gravity | Infinite | $10^{-36}$ |
| Electromagnetic | Infinite | $10^{-2}$ |
| Strong Nuclear | $>10^{-13}$ cm | 1 |
| Weak Nuclear | $<10^{-16}$ cm | $10^{-13}$ |

# 16

# e-GADS

From time to time you read that a company's business is growing "exponentially." What does this mean? Well, it's better than growing linearly. That's no answer! How many people who use this expression really know what it means? You will and that's what counts.

Linear Series: 2, 4, 6, 8, 10, 12, . . .
Geometric Series: 2, 4, 8, 16, 32, 64, . . .

Consider two sequences of numbers. A linear series grows arithmetically, by addition. In the above example, successive numbers are formed by adding two to the previous number. An exponential series grows geometrically, by multiplication. In the above example, each number is two-times the previous one. In business and finance, we prefer exponential growth because it is faster. So let's focus on it.

| Compounded Value of $1000 at 10% Annual Interest | | | | |
|---|---|---|---|---|
| No. of Years | Compounded Annually | Compounded Quarterly | Compounded Daily | Compounded Continuously |
| 1 | $1100.00 | $1103.81 | $1105.16 | $1105.17 |
| 5 | $1610.51 | $1638.62 | $1648.61 | $1648.72 |
| 10 | $2593.74 | $2685.06 | $2717.91 | $2718.28 |

If you have a savings account, the bank pays interest at a specified annual rate compounded at specific intervals. For ease of discussion, assume the rate is 10% per annum and the amount of your deposit is $1000. If compounded annually, the bank will credit your account with $100 at the end of the first year, 10% of $1100 at the end of the second year, etc.

After ten years, you would have $2593.74. A better deal is quarterly compounding, in which case you are credited every three months with 1/4 of 10% of the account balance. This increases your return because you receive interest on the interest paid during the year. These are examples of exponential growth. Most banks today compound interest daily. You can see the improved results in the table. It also shows the slightly better results of continuous compounding.

Now, let's define a new term. Refer again to the table. When interest compounds daily, total interest the first year is $105.16, or 10.516% of the opening balance. We call this the "yield." It tells how much you will actually earn if all funds remain on deposit throughout the year. Check a bank's ads. They always quote both the rate and the yield. Besides giving you a higher return, daily compounding has the advantage of your receiving interest on deposits held in an account for short periods of time, as in checking accounts.

Let's see how we arrive at our table. To calculate end-of-year value, we first compute the quantity one plus the annual interest rate (expressed as a decimal) divided by the number of times it compounds during the year. We multiply the account balance by this quantity each time interest is paid as shown.

### $1000 at 10% Interest, Compounded for One Year

$$\text{Quarterly: } \$1000 \times \left(1 + \frac{0.10}{4}\right)^4 = \$1103.81$$

$$\text{Daily: } \$1000 \times \left(1 + \frac{0.10}{365}\right)^{365} = \$1105.16$$

$$"n" \text{ Times: } \$1000 \times \left(1 + \frac{0.10}{n}\right)^n = \text{Depends on } "n"$$

$$\text{Continuously: } \$1000 \times e^{0.10} = \$1105.17$$

Here enters the transcendental number $e$. When interest compounds continuously for one year, the multiplier is $e^I$, i.e., $e$ raised to the power of the interest $I$ expressed as a decimal – in this example, $e^{0.10}$. Now, let's define $e$. From the above formula for $n$-times, we define a compounding situation where the annual interest is 100% ($I = 1$), and the number of compoundings is without bound:

$$e = \lim_{n \to \infty}(1+1/n)^n = 2.71828182884590452...$$

So, $1 deposited for one year at 100% per annum and compounded continuously grows to a value equal to $e$. This is the maximum growth possible at that rate, even if we compound every millisecond or microsecond. What about other interest rates? All right, the quantity $(1 + I/n)^n$ approaches $e^I$, as $n$ becomes larger and larger. This formula holds for any growth rate.

We can get even more practical. Let's incorporate time into a general formula, and see how we can quickly manipulate numbers the way bankers do. Let $P$ be the amount of money

$$P = P_0 e^{IT}$$

resulting from depositing $P_0$ at $I$ percent per annum for $T$ years. Try 10% interest ($I = 0.10$) for 10 years. The final value will be the original deposit times $e$, correct? Isn't this what our table tells us will happen if we compound $1000 continuously at 10% for 10 years? If your pocket calculator has a key for $e^x$ you can make these calculations very easily.

One more little trick. Investors like to know how long it will take to double the value of investments. They do it mentally without a fancy calculator. So can you. The method is very simple. First, we must ask to what power (interest times number of years) we must raise $e$ to obtain two. The answer is 0.7 (actually 0.69). This means that a value will double whenever $I \times T = 0.7$. How long will it take to double the principal, at 10% annual interest? Seven years, assuming no intervening income taxes. At 5%, it will take 14 years, etc. Clever, isn't it?

## ORGANIC RATE OF GROWTH

We have shown that continuous compounding produces maximum growth. Some mathematicians call this "organic" rate of growth because it is characteristic of phenomena in nature – the rate of growth at any time being proportional to the magnitude of the phenomenon itself. We won't quibble about whether or not growth is truly continuous, will we? From a percentage point of view, all conditions of compounding in our table produce essentially the same result. This gives us latitude to use the standard $e^x$ formulae for all exponential growth.

Bacteria and viruses grow in this manner. This is what makes AIDS so frightening. Currents in electrical circuits rise and fall exponentially. Place a hot casserole on your dining table and its temperature will decrease exponentially until it reaches equilibrium with room temperature. And exponential technological growth has brought us in sixty years from radio to television to handheld computers. We can modify our formula for financial growth, to obtain a more general one for organic growth and decay:

$$y = y_0 e^{\propto t} \text{ or } y = y_0 e^{-\propto t}$$

where $y_0$ is the initial value, $\propto$ the rate of growth or decay, and $t$ the elapsed time. Then, if we substitute $x$ for $\propto t$, we have an even more general form whose function we find on many calculators.

## CARBON DATING

Radioactive elements decay at exponential rates. With knowledge of the decay rate of different elements, we can calculate their half-life, i.e., the length of time it takes one-half of the original material to dissipate, and then one-half of the remainder, etc. In other words, radioactive elements will decay to 1/2, to 1/4, to 1/8, to 1/16, . . . in equal times, called their half-life.

All living organisms ingest Carbon-12 and Carbon-14 from the atmosphere. When organisms die, C-12 remains stable, but C-14 decays with a half-life of 5730 years. By measuring today's ratio and comparing it with the assumed original ratio between the two, technicians can estimate the age of human bones and other artifacts.

## MORE ON MATH

The phenomenon of *e* was discovered by Leonard Euler in the 18th century. He gave the lower-case Greek epsilon ε as its official symbol although the Roman italic *e* is commonly used in modern texts. The mystique of *e* is just as great as that of π. Both are transcendental numbers and are found throughout science and mathematics. More than 2200 years elapsed between their discoveries. Wouldn't it be exciting to someday discover another equally mysterious number?

Another interesting thing about *e* is what happens when we rewrite its exponential formula in logarithmic form.

*Leonard Euler (1707-1783), Swiss mathematician and philosopher.*

If $y = e^x$

then $x = \ln y$

The symbol 'ln' stands for "natural" logarithm to the base-of-*e*. If we plot these two functions on rectilinear coordinates, their curves are identical. For this reason, we often refer to exponential growth as "logarithmic" growth, hopefully not confusing it with "common" logarithms. You are familiar with common logarithms, i.e., determining to what power 10 must be raised to obtain a given number. Well, John Napier invented natural logarithms to tell to what power $e$ must be raised to obtain a given number. Here is an example:

$$\log 2 = 0.30;\ 10^{0.30} = 2$$

$$\ln 2 = 0.69;\ e^{0.69} = 2$$

Finally, just as we incorporated $r, \theta$ into our system of polar coordinates (see Chapter 11) to identify real positions, Euler discovered that we can write complex numbers (Chapter 8) in polar coordinate form $r, e^{i\theta}$, where

$$e^{i\theta} = \cos\theta + i\sin\theta$$

and use them to identify real positions.

How does $\cos\theta + i\sin\theta$ define the angle of a polar coordinate? It's not obvious but I can show you on a graph.

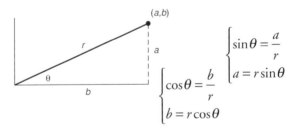

The complex polar coordinate $r, e^{i\theta}$ does indeed translate into the point $a, b$.

You will encounter $e$ in most mathematics and science courses. It is good to understand its origin and significance.

# 17
# MATHEMATICS IN ART

Mathematics is easily drawn (excuse the pun) into the field of art. Traditional art employs symmetry and balance and it shuns complexity. When your mother suggests you not wear a particular combination of clothing and accessories because it looks too "busy," she is expressing one of the fundamentals of art.

A 19th century American mathematician, George David Birkhoff, reduced this to a simple formula for expressing the typical aesthetic experience:

$$\text{Aesthetic Measure } (M) = \text{ Function of } \frac{\text{Order } (O)}{\text{Complexity } (C)}$$

Isn't this what your mother was saying? The greater the complexity, the lower the aesthetic value? Birkhoff later conceived a more ambitious goal of expanding his formula to a general mathematical theory of the fine arts.

Turning to music, we enjoy combinations of harmonically related notes and we reject combinations that are discordant. In other words, we impose mathematical limits on notes that musicians may combine. So much for generalities. Let's consider two specific ways in which we relate art and mathematics.

## PERSPECTIVE DRAWING

Successful drawings and paintings employ a vanishing point to produce a perception of depth. We call this "perspective" drawing. It depends upon the concept of a functional infinity. In both of the following drawings, the railroad tracks con-

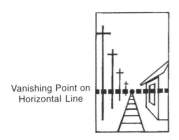
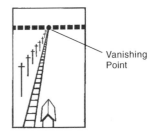

Vanishing Point on Horizontal Line

Vanishing Point

verge at a vanishing point (VP). This is interesting to me because I remember my schooldays when one of the definitions of parallel lines was that they meet at infinity. As you can see, the vanishing point is the artist's infinity. Also at this point, the telegraph poles converge to zero height.

The horizontal line passing through the VP is the "horizon line," i.e., our eye level. The higher it is in the drawing, the greater is the perceived depth. Also, the slower is the rate at which the rails and poles converge, and the less is the detail we can see in the distance.

Renaissance painters struggled for more than a century to find the mathematical scheme that would enable them to portray their three-dimensional real world on a two-dimensional canvas. Much of the credit for solving the problem goes to the $16^{th}$-century German painter and printmaker, Albrecht Dürer, for his projective geometry which gave us the vanishing point. Considering what great geometers the Greeks were, it is surprising that they never tackled this problem. Their paintings were flat. They lacked depth.

## GOLDEN SECTION

What the Greeks did not overlook, however, is that one key to artistic beauty lies in a geometric proportion called the Golden Section discovered by Euclid in the $3^{rd}$ century B.C. It is a ratio between two divisions of a line or between two dimensions of a plane figure. Let us draw two line segments,

# Mathematics in Art | 95

$$\frac{AB}{AC} = \frac{AC}{CB} \approx 1.6$$

AB, and then divide them by point C. Does your eye tell you that the lower segment with point C off center is more appealing than the upper segment in which point C lies at the midpoint? Euclid suggested that C be positioned so that the ratio of the whole segment to the longer subsegment is equal to the ratio of the longer to the shorter subsegment – i.e., AB/AC = AC/CB. We call this the Golden Ratio. Or, we say that C divides AB into the Golden Section. The value of the golden ratio is:

$$1/2 \times \left(1 + \sqrt{5}\right) = 1.61803...$$

This ratio also appears in the Golden Rectangle. I trust you will agree that a rectangle is more appealing than a square. This is optimized when length and width bear a ratio of 1.6. Human tastes have not changed over the centuries.

$$\frac{L}{W} \approx 1.6$$

Square      Golden Rectangle

The Parthenon, one of Greece's most famous temples, can be framed by a golden rectangle, as can many smaller areas inside. The next time you visit your inner city, notice how many old banks and government buildings have similar designs. Agreed, the golden rectangle does not ensure good art, but it certainly enhances it.

Look for the golden rectangle in your home. Many paintings are framed by it. Check the L/W ratio of 3×5 and 5×8 index cards, 35mm slides, photographs, and even books.

Referring to Chapter 8, review the occurrence of Fibonacci numbers in nature. The ratio of successive Fibonacci numbers (excluding the first two) is amazingly close to 1.6. Does this explain why the sunflower and the nautilus invoke beauty in our minds?

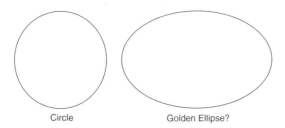

Circle        Golden Ellipse?

A goal of this book is to stimulate our thinking. It seems that we should be able to extend the golden ratio to other geometric figures. An obvious one is the ellipse. It relates to a circle in the same manner as a rectangle relates to a square, and it occurs naturally in the motion of the planets. Is an ellipse (particularly one framed by a golden rectangle) more appealing than a circle? I myself say no. Then, I think "Perhaps that is why I seldom see an oval table or mirror or painting." This disturbs me. Which is more appealing to you? Or to your friends? Can you think of other ways to test the golden ratio?

# 18
# GO GAUSSIAN

One of the more interesting discoveries in the history of science is the uniform manner in which things are naturally distributed. We are speaking of the "bell-shaped" curve, or the "normal," "Gaussian" or "random" distribution. It bears these many names, in particular that of Carl Friedrich Gauss, a 19th century German whom we call the Prince of Mathematicians.

The Gaussian curve shows the tendency of data to be distributed uniformly about a central, or average, value. We might be studying girls' heights, men's weights, factory wages, student IQ, insurance claims, scores on college entrance examinations or just about any other characteristic you can name.

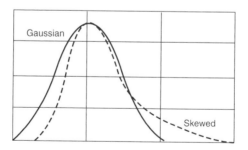

If we take enough samples, we expect to find a Gaussian distribution. However, outside events can destroy the natural pattern of events. We say that such data are "biased." For example, if a famine takes the lives of a large number of babies and children, it causes a population's age distribution to be "skewed" toward an abnormally large number of older per-

Karl Friedrich Gauss (1777-1855), German mathematician and astronomer.

sons. We show this effect on the graph. Can you visualize how an influenza epidemic that takes the lives of many older people would cause a skew in the opposite direction?

Also, in order to obtain a Gaussian distribution, all data values must be possible. If values have an upper or lower limit, the distribution is said to be "truncated." This explains why school entrance examinations are designed to be difficult enough that no one will score 100%. A perfect score would truncate the distribution, would it not? By the same token, so would a score of zero. However, considering one's fortune in answering true-or-false and multiple-choice questions, we would not expect even the poorest student to score a zero. Would you agree that a score of zero could only be achieved by a student capable of a perfect score?

Bias also occurs if different populations are mixed. Can you imagine the dual peaks you would get if you combined the height distributions of grade-school girls and adult men, or of factory wages in Russia and the United States? We call this a "bimodal" distribution.

## GAUSSIAN DISTRIBUTION

We shall discuss only the Gaussian distribution. You may ask what use is made of this phenomenon. And you should because you will be involved with it in one way or another throughout your life.

A good example is automobile insurance. The premium you will pay is based upon the insurance company's expectation of claims on your behalf. When you are young, the company expects that you will have greater claims than your parents, so they will charge you a higher premium. Unless they rate your two groups separately, they would have a bimodal distribution, wouldn't they? The shape of the two distributions may be the same, but the central value is higher for young persons. Similarly, the expectation is higher for urban residents than for rural residents.

If we take precautions to avoid bias in our data (i.e., achieve a Gaussian distribution), the laws of statistics make it very easy to make predictions about an entire universe, based upon samples from just a small portion of that universe. In other words, we can sample a relatively small number of weights, fuel consumptions, or insurance claims, and be able to accurately predict the occurrence of all values. Amazingly, this only requires arithmetic.

Let's continue our insurance example. The first step is to compute the average amount of claims for each group. This simply requires adding all sample values, $x_N$, and then dividing by the number of samples, $N$:

$$\text{Average}\,(\bar{x}) = \frac{\sum x_N}{N}$$

In statistics we use the $x$-bar notation for the average, and we use the upper-case Greek sigma $\Sigma$ to symbolize the summation of all values.

Now we introduce a new consideration, one that is a little complicated at first. The average value tells us nothing about how widely or how narrowly our values are dispersed about the average. For example, the tallest girl in an age group might be only 15% above the average, but the highest insurance claim might be thousands of times greater. Both groups follow the normal distribution, but you would never recognize it if you graphed the data in the traditional manner. This uncovers the necessity of somehow accounting for the degree of "dispersion."

Mathematicians researched this need, without benefit of computers, and found an easy solution. All we have to do is analyze our data in terms of its "standard deviation" from the average. We designate standard deviation by the lower-case Greek sigma:

$$\text{Standard Deviation } (\sigma) = \sqrt{\frac{\sum (\bar{x} - x_N)^2}{N}}$$

The formula tells us exactly what to do: 1) Subtract the first data value from the average value, square the difference, and store it, 2) Repeat for all data values, 3) Add all stored values, divide by their total number, and take the square root. Agreed? That isn't difficult to do.

Now, what does the standard deviation do for us? It's the key to the whole process! Statistical tables tell us the percent of a population that falls within one standard deviation, or any multiple thereof, of the average value. Commonly used values are shown.

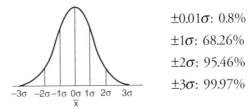

±0.01σ: 0.8%
±1σ: 68.26%
±2σ: 95.46%
±3σ: 99.97%

This "normalization" process makes our data fit the standard bell-shaped curve, irrespective of the degree of dispersion, so long as the data are random. To make it easier to read our graphs, we usually replace the average and sigma values with their corresponding data values. The vertical axis can be interpreted as either probability or number of occurrences.

Make sure you understand this process thoroughly. You will never regret the time it may take. Go back to the example of girls' heights. As long as you select the girls at random, you can expect their heights to follow the bell-shaped curve. This means you can use the graph above as a grid for plotting your distribution. You enter the average height in place of $0\sigma$, the average plus the standard deviation in place of $1\sigma$, etc. The vertical axis shows the probability of occurrence of any height.

There is still the matter of the number of samples. Whenever we evaluate just a portion of a population, we must live with the fact that our predictions for the entire population are not perfect. In other words, a subsequent evaluation using different members also selected at random will produce somewhat different results – even though the population itself remains unchanged. We must accept this risk because it is usually impossible or uneconomical to evaluate every member of a population.

What we do is quantify the error and choose a large enough sample so that the magnitude of our error is acceptable. As you might already suspect, the error also follows a normal distribution, dependent upon sample size. We calculate the "standard error" of the average:

$$\text{Standard Error} = \frac{\sigma}{\sqrt{N}}$$

In 99.97% of independent evaluations, the average will be in error by less than three standard errors. Assume in our example that $\sigma = 6$ inches and $N = 200$. The standard error is

0.42 inches, correct? So, three standard errors = 1.26 inches. This means that we are 99.97% confident that the average height calculated from a random sample of 200 girls will be in error by less than 1.26 inches from the true average. Read that again. Do you want greater accuracy? Increase the sample size.

## REAL-WORLD EXAMPLES

Statistical tables tell us what percentage of a population lies between any two values of standard deviation. A clothing manufacturer can use this information to define garment sizes and production quantities for different age groups, sexes and ethnic groups. I would be surprised if apparel industry organizations don't provide this anthropomorphic data as a service to their members.

The area under the Gaussian curve totals 100% of all the data. In our insurance example, this represents the company's total expected claims each year for each population group. Can you imagine the difficulty an insurance company would have in setting rates, ensuring a profit, and being price competitive, without benefit of this type of analysis?

Manufacturers use statistical methods to monitor product quality. They cannot afford 100% uniformity in every component. Nor can they afford not to have a reliable statistical system that alerts them to any trend toward unacceptable quality or any abrupt change in quality.

Aircraft engineers anaylze vibration data to protect against the occurrence of extremely high values, however low their probability, that may cause structural failure. They know the sigma limit that will make their airplanes superbly safe, but not burdened with excessive weight or cost. That is not to say that failures will never occur. In fact, it says there may be failures, but the prospect is extremely small – and predictable. That is what statistical analysis is all about.

# 19
# LET'S TAKE A POLL

We continually infer information about the opinions of an entire population, based upon questioning a very small percentage of that population. When polls appear as news articles in the press, there is usually an accompanying paragraph that describes the accuracy of the poll and how it was conducted.

What percentage of TV audiences watch ABC, CBS, CNN or NBC prime time news? What use is made of this information? Well, some people can be influenced to switch to the show that claims the most viewers. And the more viewers, the higher the advertising rates. It's as simple as that.

What percentage of the population watches major sports events? The reason is similar. Advertising rates for special events are based upon a network reaching a guaranteed number of viewers. Super Bowl XXIV was such a blowout in the first half that many viewers didn't watch the second half. Viewership was low for the 1988 Summer Olympics in Korea because events took place in the middle of the night our time and we knew the results before the tapes could be rebroadcast. Networks also have a viewership problem when the World Series is played between teams which lack a national audience. Advertisers are compensated if networks do not deliver as many viewers as they guaranteed.

There are also polls of taste tests, brand name recognition, product name and many other aspects of sales and marketing. Our examples could go on and on. But they all have one thing in common — the need to predict the public's actions or opinions and to do it within a reasonable degree of accuracy.

This sounds like a big job. But is it?

Let's create an example. Assume there has suddenly been a series of airplane crashes. Passenger bookings are down because people have become fearful of flying! The airlines have frantically developed a public relations campaign to restore confidence in air travel. This is very expensive. They must continually track the effectiveness of the campaign, adjust it if necessary and conclude it as soon as possible. Each element of the campaign is crucial.

You are responsible for conducting a telephone poll of 1500 adults selected at random by a computer. You find that the number of persons fearful of flying has increased alarmingly to 45% of those interviewed. We call this sample portion ($\hat{p}$ = 45%) a "statistic," i.e., an estimated 45% of the adult population in the United States is fearful of air travel. The real percentage (which we will never know because we cannot poll all 160 million adults in the country) is called a "parameter." The problem is that if you were to repeat the poll immediately, you would be calling different people and would get a different percentage. Guess what? If you were to repeat your poll a large number of times, your many statistics would follow a normal distribution about the parameter. Here we go again.

In the previous chapter, we studied distributions in which real values varied widely about a central value. Here we have a reverse situation. There is only one real value, the true percentage of those fearful of flying. Dispersed in a Gaussian distribution about it are the different estimates we would get if we repeated our poll many times. If we know the standard deviation of these estimates, we can predict their reliability.

As shown on the following curve, the standard deviation $\sigma$ is a function of the size $N$ of our sample and the probability ($p$ in %) of the real value. But we don't know the real value. If we did, we wouldn't be taking the poll, would we? Fortunately,

# 21
# More on Averages

We explored in Chapter 10 the various ways in which we characterize numeric data. Most prominent in our daily life, of course, is the arithmetic mean. What is the arithmetic mean of +8, −2, +2 and −8? "Zero," you say. Can you think of a situation where we might assign a different characterization?

## RECTIFIED AVERAGE

Assume you are standing on a platform that is moving up and down in a repetitive manner, with its relative position in space tracing out a sine wave as time passes. For every positive half of the sine wave, there is an equal and opposite negative half. Therefore your average displacement with respect to the platform's rest position is zero – no matter how wide the excursion. But would you describe the effects of your motion

as zero? I guess not! If the movement is very small, as often happens on a boat, you might become nauseated. If the motion is large and you are holding a bottle of orange juice, it will soon be well mixed, won't it? Just as you, it knows when it is being shaken. How can we more meaningfully characterize the amplitude of this motion?

Our first step is to modify our graph of time history by taking all the negative motions and presenting them as being positive. We borrow a term from the electrical engineering

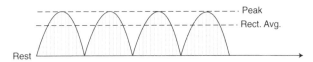

field and say we have "rectified" the waveform. What were formerly negative values are now treated as positive values. In mathematics we call them "absolute" values and symbolize them as, e.g., |8|, meaning we use its value without regard for its sign.

Next we calculate the arithmetic mean of all the points on the curve and report it as the "rectified" average, often loosely referred to as simply the average. It tells the average excursion of your motion without respect to direction from the rest position of the platform. This is a better indicator of the effects of motion than is the true average of zero, isn't it? Since this motion is sinusoidal, you can also calculate your peak position which is 1.57-times the average (actually $\pi/2$ times), or the arithmetic average which is 0.636 times ($2/\pi$ times) the peak.

## RMS VALUE

Our electrical engineering friends had a different problem. When they applied a sinusoidal voltage to a resistor, the heat energy generated by the current passing through the resistor was proportional not to the rectified average value of the voltage, but to 1.11 times (actually $\pi/2\sqrt{2}$ times) the average. This gave them the "energy-related average" for a sine wave. But how do we characterize the energy of signals whose formula or waveform we do not know? I'll make a long story short. The energy-related value is the root-mean-square (RMS) value – the same parameter introduced by its statistical name of standard deviation in Chapter 18. Another coincidence of nature!

In the real world we almost always deal with unknown

waveforms in the form of sound, vibration, strain, electrical noise, pressure fluctuations and other dynamic phenomena, so it is the RMS value that we measure most often because it relates to the energy of our signals. Fortunately it is easy to measure with today's instrumentation. We seldom if ever measure the rectified average because it does not correlate well with real-life effects.

## PEAK VALUES

Now let's climb aboard a trampoline in your room and have some fun. Your parent soon hears all the commotion and shouts, "Don't jump too high! You'll hit your head on the ceiling and hurt yourself!" What in the world does that have to do with the average or RMS value? Well, very little. The

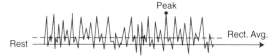

only thing we can say for certain is that it has a greater amplitude. And by a big margin. Correct? In cases of impulsive phenomena, such as sonic boom, earthquakes and mechanical shock, we always measure peak values because they are the most damaging to structures. In special cases, such as evaluating shaft unbalance in large rotating machinery, we measure the peak-to-peak excursion between positive and negative peaks.

## AMPLITUDE DISTRIBUTION

Returning to the phenomena that we characterize by RMS value, we classify most of them as "random," meaning that the signal is not repetitious, that it contains a broad band of frequencies, and that its instantaneous amplitude varies unpredictably.

We showed that the RMS value characterizes the energy

content of the signal. But what about instantaneous amplitudes? Peak values can be critical! With sine waves, it was simple. Tell me the frequency, the peak or average or RMS value, plus a time reference and I can tell you its amplitude at any time past or future. Not so with random signals. We seem to know nothing about their instantaneous amplitudes. Or do we? Well, once again nature comes to our rescue! It so happens that if the signal is truly random its amplitude variation will follow a Gaussian distribution about its central position. Isn't that convenient!

We saw in Chapter 18 that if we know the standard deviation of a Gaussian distribution we can predict the long-term probability of occurrence of any individual value. Analytical instruments today can sample a signal repeatedly and either confirm that its amplitude distribution is Gaussian or else show us the actual distribution. In either case we can then predict the probability of occurrence of extreme values for individual signals. No matter where we turn, we seem to find examples of nature's famous bell-shaped curve.

## 22

# ZANY WORLD OF WEIGHTS AND MEASURES

With due respect to scientists of the past 300 years, I must say that they did some zany things as they developed our systems of weights and measures. We'll explore some serious and lighter aspects of their efforts with our most common measurements.

After the French revolution, Talleyrand (best known as a masterful politician and diplomat) proposed that the French Academy of Science reform France's system of weights and measures. A blue-ribbon committee was appointed. Although invited to participate, Britain and the United States declined on the basis that reform was impractical. To this day, they refuse to fully adopt the world's metric system.

### LENGTH

Talleyrand's primary concern was the unit of length, because the Parisian unit differed by as much as 4% from measures used in other regions of France – not a very good deal for commerce. The academy set two goals. The length should be practical for everyday use and it should be traceable to a physical constant.

Someone came up with the bright idea that one ten-millionth of a quadrant of Earth's polar circumference would be a suitable length. You can divide it out, starting with the circumference being about 24,000 miles. My calculation says

3.168 feet by this measure. It would later be named the meter, from the Greek word metron, for measure. They chose to survey a sector of about one-ninth of a quadrant (10 degrees of latitude) between Dunkerque on the English Channel and Barcelona, Spain. This task took from 1792 to 1799, and it gave them a standard meter that differed by only 3 parts in 10,000 from their estimates.

Big deal! They spent seven years establishing a standard length, which is a little longer than an arm's length. But they had the satisfaction of "knowing" that it was an even division of Earth's polar circumference. Modern measurements, however, show their standard to be short by about 2 parts in 10,000.

In the United States, we took a more cost-effective approach. We accepted the British yard as our standard. It's about 91% of a meter. When I was a child and my mother bought "yardage" goods to make a tablecloth or curtains, the sales clerk measured off each yard as the distance between her nose and the end of her outstretched arm and fingers. Perhaps this anatomic relationship was the basis for the yard. The yard is also the typical length of a man's stride, so it is easy to pace off yardage on the ground. Once agreed upon, the yard was just as valid a standard as the meter. And it didn't take seven years to define, although it probably would today!

From the late 18th century until 1960, the standard meter length was "preserved" as the distance between two engraved lines on a special platinum-iridium bar stored in a vault outside Paris. Our yard was similarly preserved. After painstaking comparisons between their bar and ours, the yard was eventually redefined as 0.91440183 meter.

Time marches on! Today we can measure lengths more accurately by using light waves. In 1960 the 11th General Conference on Weights and Measures redefined the meter to be the length equal to 1,650,763.73 wavelengths in a vacuum of a specific radiation from the Krypton-86 atom.

Why is the meter the overwhelming global preference? The answer could not be more simple. It is found, not in the meter's length, but in how it is divided. The French loved the decimal system. They divided the meter into ten equal divisions called decimeters, the decimeter into ten centimeters, the centimeter into ten millimeters, etc. Every length they measure is scaled as a decimal portion of a meter.

On the other hand, we went bonkers! We observed that the length of a man's foot was about one-third of his stride, so we created the unit of 1 foot = 1/3 yard. Many of us assume that the foot is our standard unit of measurement. And why not? About the only time we encounter the yard is on a football field or golf course.

Having divided the yard into 3 feet, we proceeded to divide the foot into 12 inches and then used the binary system to divide the inch into halves, quarters, $8^{ths}$, $16^{ths}$, $32^{nds}$ and $64^{ths}$. Wow! Figure it out yourself. The divisions for our standard yard are 1/3, 1/36, 1/72, 1/144, 1/288, 1/576, 1/1152 and 1/2304. Is it any wonder the world did not follow us?

Is it also any wonder that we treat the foot as if it were our standard. At least we rid ourselves of that undesirable division by three. Many in our engineering world say "We will have no part even of the binary division of the inch. We will apply the decimal system to our inch." We call their ruler the "engineer's scale."

## WEIGHT

Another custom of the metric system is that, wherever possible, units not fully traceable to a physical constant are defined, at least in part, by units that are. For example, there is no weight in nature that the French could find to use as a standard. So they defined the gram as the weight of pure water at 4° C (point of maximum density) in a cube whose sides are

1/100 of a meter, i.e., 1 cc of pure water at 4° C. Now, here is something interesting. Although the gram is the metric standard, the International System (SI) of Units adopted the kilogram as its basic unit.

In the United States we have the avoirdupois pound divided into 16 ounces, properly called avoirdupois ounces to distinguish them from apothecary ounces of which there are 12 to an apothecary pound. Then we introduced another confusion factor by dividing the quart into 32 fluid ounces. The examples of zaniness go on and on.

## YOU CAN'T WIN THEM ALL

Try as they did, the French could not sell the public or anyone elso on a decimalized clock. Clock reform was tabled by law in 1795 and remains tabled today. The only concession to change is the 24-hour military clock which eliminates the ambiguity of AM and PM designations.

Think about it. Decimalizing the hour sounds like a good idea. But there are 24 hours in a day. Should the day be decimalized also? The French Academy proposed just that: 1 deciday = 2.4 hours, 1 centiday = 0.24 hours, etc. They could go no further up the ladder of time because the day and the year are determined astronomically by the rotation of Earth about its axis and by Earth's rotation around the sun, and they are not in decimal ratio to the day. No wonder the idea was rejected.

A few punches did get through. In timing athletic competitions we decimalize the elapsed seconds. And we use scientific notation to define extremely long and short periods of time.

## WHAT IS THE TEMPERATURE?

This one really threw scientists for a loop. They had built their beautiful system of measurements referenced in whole or in part to physical constants, but they were stymied when they came to temperature. There was nothing they could borrow from nature to measure heat content.

The first person of consequence to explore a methodology was a Danish astronomer, Ole Roemer, in the early 1700s. He was no lightweight. He is noted for discovering that light travels at a finite velocity, not instantaneously as previously believed. He had the help of prior thinkers (including Newton) who had suggested that an arbitrary temperature scale could be established by relating it to the freezing and boiling points of water. In a way, these were excellent choices because temperature is constant during the freezing or boiling process. This would simplify calibration. Roemer was also interested in another constant – the lowest temperature attainable. In this he was very prophetic.

He built an alcohol-filled glass thermometer tube and proceeded to experiment with various proportions of salt, water and ice until he achieved a minimum temperature. Now he was ready to calibrate. Being an astronomer, he was used to the sexagesimal (1/60) system. So he marked a zero on his scale for the lowest temperature he attained and 60 for the point at which water boiled and divided the scale into 60 equal divisions. In freezing water, the reading was 7.5 degrees – degrees "Roemer," to be correct, indicating that the scale was arbitrary.

In 1708 while still perfecting his calibrations he received a 22-year-old visitor, Daniel Gabriel Fahrenheit, of Holland, who was a manufacturer of scientific instruments. When he returned home, Fahrenheit set about to improve Roemer's thermometer. His chief contribution was the use of mercury for the fill. He also achieved a lower minimum temperature by adding sal-ammoniac to his mixture. Then, judging that a 60-degree range was not sufficiently sensitive for everyday variations, he changed the range to 0 to 240 (later corrected to 0 to 212). The freezing point of water occurred at 32 degrees. This

Fahrenheit scale is still in use almost 300 years later, but with sensing techniques that enable us to measure much higher and, yes, much lower temperatures.

A few years later, from the Swedish university town of Uppsala, came an astronomer named Anders Celsius who decided that the range from freezing to boiling should be divided into 100 divisions. This made him the darling of the metric crowd. But he discarded the idea of identifying a minimum temperature. He assigned boiling water the reading of zero and freezing water a reading of 100. How strange. The greater the heat the lower the reading! Shortly before his untimely death in 1744 his peers properly inverted the scale. The name assigned was degree centigrade, meaning hundred-degree. Not until 1948 was the name changed to degree Celsius. This was a great honor, but my research turned up no other contribution by Celsius than the 0 to 100 scale. Ironically, his deletion of a minimum temperature showed he did not have as good a grasp of the physics of heat as did Roemer and Fahrenheit.

From a scientific point of view, all of their scales have deficiencies. A minor one is that extension to negative readings can give the false impression of negative heat. But we know that "cold" does not exist; only "heat" exists. The major defect is that the scales do not have a proper zero – not by a long shot. The ratio between the boiling and freezing points of water is 6.6 on the Fahrenheit scale and infinity (100/0) on the Celsius scale. How about that? It means that one or both of the scales are not proportional to heat content. They are not absolute values. Sooner or later someone would discover the conditions for "absolute" zero.

The foundations were laid in the mid-19th century. The best known scientist in this effort was William Thomson Kelvin, better known by his later title of Lord Kelvin. By then scientists knew that heat is caused by the friction of atomic

|  | Fahrenheit | Rankine | Celsius | Kelvin |
|---|---|---|---|---|
| Boiling Water | 212° | 671.67° | 100° | 373.16 |
| Freezing Water | 32° | 491.67° | 0° | 273.15 |
| Salt Solution | 0° | 459.67° | −17.78° | 255.37 |
| Absolute Zero | −459.67° | 0° | −273.15° | 0 |

and molecular motion. When this motion ceases, nature's zero temperature exists. This is the physical constant Talleyrand's French Academy would loved to have had. By extrapolating the Celsius scale, absolute zero was discovered to fall at a temperature of −273.15° C. Scientists immediately added 273.15 to all readings on the Celsius scale and gave the new scale the name Kelvin. Similarly, from the corresponding magic number of −459.67° F, they created the Rankine scale, after the Scottish scientist, William J.M. Rankine. These new "thermodynamic" scales give us absolute readings. A body with twice the Kelvin or Rankine reading of another body is twice as hot. How much greater is the heat content of boiling water than freezing water? Does it matter whether you use the Rankine or the Kelvin scale to obtain the answer?

The ending to this story came in 1967 when the degree Kelvin was renamed the kelvin and elevated to the position of a defined unit in the S.I. system, putting it on a par with the meter, kilogram, ampere, etc.

## WHERE WE STAND TODAY

The push for metrification has not been a resounding success in the United States. Global industries responded, of course. If a factory in Detroit builds automotive engines for use in both Europe and America, it cannot afford to operate two different lines. It must metrify. The same is true for aircraft and industrial machinery.

Where commodities are involved, we only pay lip service. Check the food containers in your pantry. I have a jar of peanuts labeled 1 lb (453 grams) and a jar of popcorn labeled 2 lbs (907 grams). One pound is actually 453.6 grams, so each manufacturer merely rounded off to the next lowest whole gram. This is called 'soft' metrification. We do the same with highway distances and speed limits and a host of other things.

We resist metrification because we cannot think in terms of metric units. Is your school teaching you to think in terms of meters and kilograms? How long might it take to accomplish this in our society? Or, is it even necessary?

## 23
# Secrets of the Cone

We frequently use a popular geometric design to hold ice cream, popcorn and ice slushes. It is the cone. This practice is not unique to us. Farmers in developing countries roll sheets of newspaper into the shape of cones to hold the vegetables and grain they sell in street markets. When I encounter these scenes, I envision farmers of long ago fashioning similar holders from large vegetable leaves.

Conic shapes also occur in nature. Examine pine cones or photographs of mounds created by volcanic eruptions. And if you have a sandglass in your kitchen, observe the development of a cone shape as sand accumulates at the bottom. My dictionary defines a cone as a solid generated by rotating a right triangle about one of its legs, called its axis. We can also visualize it as an expanding circular surface emanating from a vertex.

As you know, early Greek mathematicians were fascinated by geometry. They discovered that if they sliced a cone with a plane perpendicular to its axis, the curve of intersection would be a circle. The next logical step was to study intersections with planes that were not perpendicular to the axis. Extension of these studies over the centuries led to many secrets that became a part of modern technology.

The person credited with developing the geometry of the cone is Apollonius of Perga. He followed Euclid, developer of the geometry of points and straight lines in a plane, by about seventy years. His first step was to tilt the plane so it was no longer perpendicular to the axis, but not tilted so much as to be parallel with one of the cone's lines of generation. Can you

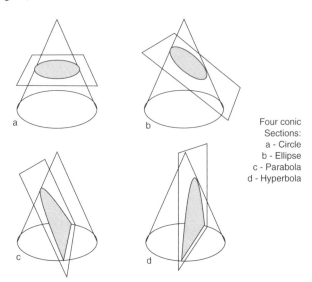

Four conic Sections:
a - Circle
b - Ellipse
c - Parabola
d - Hyperbola

visualize the curve of intersection as being an ellipse? If we continue to tilt the plane until it is parallel to one of the lines of generation, the intersection is an open-ended curve called a parabola. Tilt it further and the open-ended curve is a hyperbola. The Greeks determined that these are the only possible curves of intersection. We call them "conic sections."

About 500 years after Apollonius, around 300 A.D., Pappus of Alexandria discovered that the conic sections can be defined in a plane by loci of a point, P, that moves so that the ratio of its distance from a fixed point, F, (focus) to its perpendicular distance to a fixed straight line, N, is a constant. As this ratio is less than, equal to, or greater than 1, the curve is an ellipse, parabola or hyperbola. We will see that there are other definitions as well, demonstrating that the one by Pappus is not unique.

## CIRCLES

Let us begin with the circle. It is a special case of an ellipse, but it is easier to define it as the locus of a point that moves so

# Secrets of the Cone

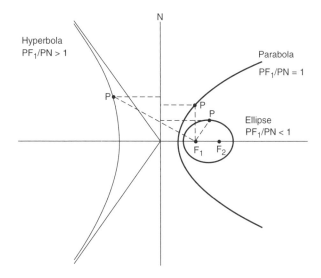

*Conic sections with focus at $F_1$.*

it is always a constant distance from a fixed point. The first application of geometry to real-world needs might well have been when an ancient civilization discovered it could revolutionize its method of transportation if craftsmen would fashion a pair of circular disks and attach them to their cargo boxes and chariots! Get it? The wheel, of course.

An important property of a circle is that it encloses a greater area than any other closed curve having the same perimeter. Compare a circle to a square. Assume a perimeter ($2\pi r$) of 10 feet. A circle would have an area ($\pi r^2$) of 7.96 square feet vs. 6.25 square feet for a square. Correct? For any perimeter, a circle encloses more than 27% greater area than a square (ratio is actually $4/\pi$). The advantage over a rectangle is even greater. These are significant differences! Perhaps that is why water pipes are circular. This property is also very important to architects, isn't it? Circular buildings yield the most floor space for a given amount of construction material.

We can also say that a circle is the shortest possible curve that encloses a given area. The ancients knew this. The Phoenician princess Dido obtained a grant of as much land as she could enclose with an ox-hide. She cleverly cut the hide into long thin strips and staked out a semicircle of land along the North African coast. The land became the State of Carthage and she the Queen.

If we take another step and rotate a circle about its diameter, we get a sphere. Does it surprise you that a sphere is the smallest surface area that contains a given volume? Soap bubbles are spherical because the soap film's natural tendency is to assume a state of lowest surface energy. Any other closed shape of the same volume would have a higher surface energy. Plato called the circle the most perfect figure and the sphere the most perfect solid.

## ELLIPSES

The conic sections graphic illustrates the generation of an ellipse, according to Pappus. The ratio shown between the distance from P to $F_1$ and the perpendicular distance from P to line N is 1/2. Observe the symmetry of the ellipse and the position of $F_1$. Does it suggest the possibility of a second focal point at $F_2$? My dictionary defines an ellipse as the locus of a point, P, that moves so that the sum of its distances from two fixed points, $F_1$ and $F_2$, is a constant. This definition was used in Chapter 6. The greater the distance between the foci, the greater the eccentricity of the ellipse. In other words, the more oval it is. Conversely, the closer the foci, the more circular it is. When the foci are coincident, the locus is a circle.

It is interesting to note that when you look obliquely at a circle, the front-to-back dimension is foreshortened and what you actually perceive is an ellipse. From a distance, a round table or the rim of your drinking cup, appears to be elliptical. Might we call these virtual ellipses.

## PARABOLAS

The parabola is the easiest of the conics to plot using Pappus' definition – the distances from a point and from a line being equal. The parabola is of tremendous practical importance in many areas of technology. Let's begin with Galileo. One of his many important discoveries evolved from his study of the path of projectiles. That a cannon ball moved forward while it also rose and then fell was obvious to everyone, but no one really understood the mechanics of the motions. Galileo showed that the trajectory could be resolved into two simultaneous motions – a horizontal motion having constant velocity and a vertical motion that conforms to the law of falling bodies. He also showed that the combination of the motions is the path of a parabola. He then developed his Superposition Principle which states that component motions can be analyzed separately and then combined to yield a net result. With this knowledge, the aiming of cannons advanced from an art to a science. Had airplanes been in existence, Galileo would have told us that the path of a projectile dropped from an airplane is that of half a parabola. We use this knowledge when aircraft attack forest fires with water bombs, the water being released at the precise moment that will enable its parabolic trajectory to intercept the fire area. Similarly, an object dropped from a moving ship falls in a parabolic path.

The parabola is extremely useful when we rotate it about its axis to form a surface. In the time of Sir Isaac Newton, telescopes employed a spherical lens to focus images. The problem was that images were fuzzy (in the eye this affliction is called astigmatism) because light rays that arrived near the edge of the lens did not focus at the same point as rays that arrived near its center. Newton redesigned his telescope and replaced the lens with a mirror. He chose a parabolic surface for the mirror because he knew that a parabola would precisely focus all incoming light rays that were parallel to its axis. This was the world's first reflector-type telescope. In 1781, William Herschel, an organist in Bath, England, was using his homemade reflecting telescope when he discovered Uranus – the first planet discovered in modern times. Today we apply the same parabolic shape to satellite receivers and TV dishes in order to focus incoming electronic signals upon a sensor located at their focal point.

The process works in reverse as well. While researching this chapter, I came upon a portable space heater that employed a parabolic reflector with a heating coil at its focal point. It was labeled a "parabolic" heater and bore the descriptive trade name of Heat Dish. You guessed it, the reflector directed the heat energy in a forward direction to make it more efficient in small areas of a home or office. In another example, flashlights and automobile headlights aim light by employing a parabolic reflector behind the light bulb. Of course, you wouldn't use a parabolic reflector if you wanted to disperse the energy, would you?

It is interesting to note that a parabola results from a unique condition – a plane of intersection being parallel to a cone's line of generation or a point moving so that its distance from a focus equals its distance from a straight line. Either side of this unique boundary, an infinite number of conditions produce ellipses and hyperbolas.

## HYPERBOLAS

When we plot a hyperbola, the focus lies outside the curve. If we construct a second hyperbola (not shown) having its focus at a corresponding location inside the first hyperbola, we have a pair of hyperbolas that are mirror images of each other. We obtain this pair of curves when a cutting plane intersects a pair of identical coaxial cones joined at their vertices. Also note that hyperbolas are "asymptotic" to a pair of boundary lines, meaning that the curves approach but never intersect these lines, called asymptotes.

The three conic shapes are among many shapes used to optimize the performance of acoustical horns. The moving part is usually a cone shape. In fact, we use the word 'cone' to identify the part of the loudspeaker that causes the air to move.

## PATHS ACROSS THE UNIVERSE

Imagine a cannon being located in space ninety-three million miles from a star like our sun, the same distance as Earth is from the sun, with no spatial objects to cause gravitational interference. Also imagine that we fire the cannon in a direction 90° to the direct path to the star. With these initial conditions, let us see what happens as the exit velocity of the cannonball varies. First, assume zero velocity, as if it were merely dropped. You guessed it. The cannonball would fall in a straight line to the star, with continuously increasing velocity caused by the gravitational attraction of the star.

Now, let us give the cannonball a little bit of speed. Combining its initial forward force with the gravitational attraction causes the cannonball to set out on a highly elliptical orbit, with the star located at the nearer focus. At increasingly greater exit velocities the cannonball's path becomes less elliptical until at about 18.6 miles per second it becomes a circle. Continue to increase velocity, and the path reverts to an ellipse but with the star at the further focus.

At about 68 miles per second, the cannonball will have sufficient velocity that it will break away from the attraction of the star and go off into space in a parabolic path that stretches to infinity. As it gets farther from the star, its velocity continually decreases until coming to rest at an infinite distance. Make the initial velocity even greater, and it follows the path of an hyperbola and will never come to rest (see *Bound to the Sun*), being hardly affected by the star's gravity and traveling along its asymptote.

We have described an ideal system consisting of one star and a nearly massless cannonball. In reality, of course, this condition does not exist in the Universe. Therefore the paths of stars and planets and other space objects are not perfect conics, but we think of them as such. The large masses that became stars, planets and moons had relatively low velocities when captured, so they orbit in elliptical paths. The same can be true of comets. Halley's comet is the best known example. It has a period of 75 years, having last passed our way in 1986 and is next expected in 2061. Smaller comets tend to be traveling faster when captured, so many of them follow parabolic and hyperbolic paths. They pass our way and never return.

## FURTHER RECKONING WITH NATURE

Although not directly related to the conic sections, there is a similarly important curve in nature. Let's pick up the story during the lifetime of Galileo as he pondered the efficiencies of nature. "I think that no one believes that swimming or flying can be accomplished in a manner simpler or easier than that instinctively employed by fishes and birds," he wrote. "When, therefore, I observe a stone initially at rest falling from an elevated position and continually acquiring new increments of speed, why should I not believe that such increases take place in a manner which is exceedingly simple and rather obvious to everybody?" Galileo went on to prove that a body moving down an inclined plane from a given height attains a velocity independent of the slope of the plane and is the same as what it would have been had the body fallen through the same vertical height. But something must change with slope, and he found it to be the elapsed time of the fall. He then studied other paths. Contrary to one's normal assumption that a straight line is the shortest and therefore the quickest, he observed some paths that were faster. Hence the question – what is the fastest path?

About a hundred years later, in 1696, the Swiss mathematician, John Bernoulli, became so interested in this question that he posed it as a challenge to the mathematicians of Europe. It was solved by him and by his brother James, as well as by Newton, Leibnitz and others. It is known as the "Brachistochrone" problem, and it initiated a new branch of mathematics, called Calculus of Variations. The curve of quickest descent proved to be an arc of a cycloid. By now, you probably expect me to tell you that the cycloid is a very common curve. And it is. It is the locus of a fixed point on the circumference of a circle, as the circle rolls along a straight line. Substitute the word wheel for circle, and you discover that every point on the tread of your automobile tires traces out a cycloid over and over again as you drive down a highway. I recall doing a paper on the cycloid once upon a time and learning that birds dive in a cycloidic path. Get it? The faster their descent, the better their chance of catching the prey.

We solve such problems repeatedly in the modern world. For example, what vehicle shape creates the minimum air resistance and hence the lowest fuel consumption? There are also many problems in economics. Take the example of a mine with a finite supply of ore. What economic model yields the greatest overall profit? High initial output which decreases later, constant output, increasing output with time, or a combination thereof? Many improvements in design and production are attributable to optimization problems solved by the Calculus of Variations.

## WORD DERIVATION AND USAGE

Let's return to the conic sections. Pythagoras used the terms "ellipsis," "parabole" and "hyperbole" to define mathematical conditions of "less than," "equal to" or "greater than," and Apollonius adapted this nomenclature to the conic sections. The terms have come down to us as ellipse, parabola and hyperbola.

The order in the conic sections represents conditions of less than, equal to, and greater than. We saw it first when we observed that when a cone is intersected by a plane parallel to one of the cone's lines of generation, the curve of generation is a parabola. With less tilt the curve is an ellipse, greater tilt a hyperbola. When defined as the loci of a point with respect to a line, the curve is an ellipse, parabola or hyperbola accordingly as the ratio between the distance from point to line and point to focus is less than, equal to, or greater than unity. Finally, the cannonball's path progressed from ellipse to parabola to hyperbola as its initial velocity increased.

We can also note that from parabola we have the word comparable (equivalent), and from hyperbola we have hyperbole (excess). And elliptic can have the meaning of economy or smallness.

## THE HUMAN EYE

Examine how nature employs the conic shape to make the human eye more efficient. Behind the retina are about 125 million photoreceptors which connect through the optic nerve to the brain. About 94% of them are cylindrical photocells that enable us to see images in the manner of a black-and-white camera. The rest separate colors into blue, green and red so we can see images in color. These are less sensitive but use the focusing power of a conical shape to gather more light.

We have seen remarkable stories of how the conic sections pervade our world and universe. What does the occurrence of these elegant curves have to do with the results we get when a plane slices a cone? This is truly a mystery.

# 24

# LONGITUDE, OH, LONGITUDE

You are being treated to a broad sampling of successes in the math and science arena dating back to the ancient Greeks, so it is only fair to tell you about a problem that waited far too long to be solved. We are speaking of the failure until 250 years after the time of Columbus, Magellan and Balboa, to provide navigators with a reliable determination of longitude while at sea.

Eratosthenes invented the idea of a grid pattern to identify place locations on the globe, and a century later Hipparchus adorned our planet Earth with the latitude and longitude lines we still use today. The equator marks the natural dividing line between the Northern and Southern Hemispheres. The prime meridian of longitude is an arbitrary meridian that serves as a time reference for clocks throughout the world. As youngsters we spotted familiar place names on a globe and learned to identify their location by their latitude and longitude coordinates.

At the International Meridian Conference held in Washington, D.C. in 1884, the meridian passing through the site of the Royal Observatory in Greenwich, England, seven miles from the heart of London, was assigned to mark the world's time reference (Greenwich Mean Time), or 0° longitude. The world's time is kept there by an atomic clock accurate to within millionths of a second. Locations to the east or west have local times that are ahead of or behind Greenwich Mean Time (GMT). Longitude lines are numbered from 0 to 180°

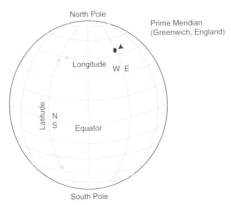

east and west, meeting at the International Date Line. Time zones around the world register a legislated number of hours ahead of or behind GMT. In airline jargon, pilots borrow the word 'Zulu' from their phonetic alphabet to designate GMT. Europe is a good location for the prime meridian because it places over the vast, unpopulated Pacific Ocean the 180° meridian where travelers gain or lose a day.

With precise, low-cost Global Positioning Systems installed today on most pleasure boats and many automobiles, we tend to forget the days when sea navigation was as much an art as a science, and a difficult art at that. This is a brief history of navigation and the role that numbers played in it.

If a sea captain ever wanted to return home from a voyage, he had to carefully chart his journey. That meant continually plotting his coordinates of latitude and longitude. Latitude was fairly straightforward, but longitude was very difficult.

## NAUTICAL MILE

Sailing enthusiasts might be offended unless we talk in terms of the nautical mile (nmi). It is an interesting unit of measurement in its own right. We divide longitude into 360 degrees. By definition, the length of one nautical mile is the distance *on the equator* represented by one minute of longitude. Therefore 60 minutes of longitude, or one degree, represents 60 nautical miles. This establishes the circumference of Earth as 60 × 360 = 21,600 nmi. This is all well and good, but it does not tell us the length of the nautical mile or the equatorial circumference, does it? To make the definition work, someone has to tell us the circumference of the equator in standard units of measurement.

For a long time mapmakers used England's value of 6080 feet for the nautical mile. However, it did not represent an even number of meters, so the international community arbitrarily changed the length of the nautical mile to 1852 meters, or 6076.1 feet. The interesting thing is that this action reduced the value long used for equatorial circumference! A nautical mile is 6076.1 feet vs. 5280 feet for statute mile, so the constant used for equatorial circumference is 24,856.77 statute miles. Earth is for all practical purposes a perfect sphere, so one degree of latitude also equals 60 nmi. Take the ratio of the number of feet in the two versions of a mile, and you will find that 1.0 nautical mile equals 1.15 statute miles. Another item of note is that sailors use the term 'knot' for a speed of one nautical mile per hour.

Also observe that one degree of longitude represents one nautical mile only on the equator. It is less at any other latitude, e.g. 52 nmi at 30° latitude and 30 nmi at 60° latitude, because the lines of longitude converge at the Poles where the circumference of the globe is zero. On the other hand, one degree of latitude equals one nautical mile at all latitudes.

## EARLY DAYS OF NAVIGATION

Nature provides valuable navigation aids to mariners. The ancients knew that the Pole Star (Polaris) remained at approximately the same position in the sky throughout each night and that it dipped toward the horizon as they sailed south. They devised simple hand-held devices to measure the Pole Star's angle with the horizon because it gave them a direct measure of latitude. These devices have come down to us today in the form of sextants. The same measurement was performed at noon when the sun was at its highest point in the sky. So by night and at one time during the day, *weather permitting*, they had a good fix on their latitude. Or did they? The problem is that as seasons change, the sun moves higher and lower in the sky. This meant they had to take the sun's seasonal factor into consideration, so their noon measurements were less reliable.

Longitude determination was a different story because as Earth revolves on its axis there is nothing about the positions of the sun or stars that relate to longitude. Oh, there were celestial events such as eclipses and phases of the moon whose dates were known, but they were far too infrequent to be of any real value.

One of Magellan's crew members reported that the great explorer spent many hours studying the problem of longitude, "but," he said, " the pilots content themselves with knowledge of the latitude and are so proud they will not hear speak of longitude." Many others felt the same way.

A mariner's only recourse was dead (from deduced, or ded, for short) reckoning, i.e., repeatedly measuring their direction, speed in the water, and time on-course and then plotting their movement on charts. For speed, they threw a traverse board astern of the ship and estimated the ship's separation distance over a short period of time as measured by a sand glass. For direction, they had compasses. Errors were cumulative, but the biggest problem was they did not know and could not compensate accurately enough anyway for winds and currents. Captains eventually did gain some knowledge of familiar areas and began to turn navigation into somewhat of an art which was guarded jealously because crews were less likely to mutiny if they didn't know where on Earth they were.

In the early 1600s, a Spanish admiral sailing from the west coast of South America discovered what are now the Solomon Islands off the coast of New Guinea. He didn't claim them because he thought they were an unmapped area of the Marquesas Islands – on the same latitude but 4000 miles to the east.

When bound from England to the West Indies they could cleverly take a southwesterly course to about 20° north latitude and then "sail the parallel" until they reached the islands. That's what Columbus did. The return voyage was more difficult, wasn't it? This narrowing of sea routes led to exploitation by pirates. By 1700, some 300 cargo ships sailed between England and the West Indies each year.

Some ships carried shore-sighting birds which were released when the captain became lost. If the birds flew off, the captain was pretty certain he could reach land by following the direction of the birds. Being lost often meant exhaustion of his supply of fresh fruits and vegetables which were rich in vitamin C. This led to the dreaded disease of scurvy. In later times English mariners countered with a supply of limes – hence the name "limeys" for British sailors.

Astronomers, including Galileo, Newton and Halley searched for ways to use the stars and planets to determine longitude. But they were unsuccessful. Unbelievably, longitude defied reliable calculation until the 18$^{th}$ century.

## CHRONOMETERS

The secret to accurate determination of longitude was not in the stars but in Earth's own "time clock." Earth revolves once every 24 hours and is divided into 360 degrees. It therefore rotates through 15° each hour, or 1° every 4 minutes. It therefore advances about 15 nautical miles of longitude per minute of time, on the equator.

They could determine noon each day by the moment when the sun was at its highest point in the sky. Now, if they had an accurate clock (chronometer) to read Greenwich Mean Time, they could easily calculate their longitude – on the basis that they were east or west of the prime meridian by 15° for each hour plus 1/4° for each minute that local noontime was ahead of or behind the prime meridian.

The problem was that accurate timepieces were not available. This is not surprising. People had little need for clocks in those days. Sunset and sunrise told them when to go to bed and when to get up. Factory whistles told them when to go to work, when to have lunch and when to go home. And church bells told them when to worship.

In fact, mechanical clocks were not invented until about 1500. They were crudely built and could not withstand the rigors of sea duty. More importantly, they were not accurate enough. The first successful chronometer designed for sea duty was built in England by John Harrison who in 1773 won the prize offered by Parliament in the Longitude Act of 1714. He succeeded in making a science of longitude determination at sea.

Time marched on. Chronometers became obsolete with the advent of radios to continually broadcast Greenwich Mean Time. Technology advanced through radio navigational aids, inertial navigation systems (precise electromechanical dead-reckoning systems) and now, satellite Global Positioning Systems. In a pinch, experienced navigators today can use their sextant, navigation almanac and for a chronometer all they need is a cheap digital wrist watch.

# 25
# EQUATIONS – OUR BEST FRIENDS

In the world of business and science we present data in the form of equations, graphs and tables. There is a great difference in the amount of information each reveals. Equations, which most students hate, are actually our best friends because they reveal the most information and present it in the most concise format. If I limit myself to treating:

$$c = \pi d$$

as a routine formula, I miss the whole picture. The only thing I learn is how to calculate the circumference of a circle when I know its diameter, or vice versa. But I learn much more when I treat it as a statement of *relationships*. With a single glance I observe that circumference and diameter both appear to the first power and that they are the only variables. This tells me there is a linear relationship between them, i.e., if one changes, the other changes by the same percentage. And since no limits are indicated, the relationship holds no matter how large or small the diameter might be.

When there are more than two variables, as is usually the case, equations reveal the tradeoffs available as we manipulate parameters to accomplish our goals. An elementary equation in structural mechanics defines the resonance of spring-mass systems:

$$\text{Resonance Frequency} = \frac{1}{2\pi}\sqrt{\frac{\text{Stiffness}}{\text{Mass}}}$$

Motion will be greatly amplified if excitation forces approach the structure's resonance frequency. To prevent damage, structures are designed so their resonance is considerably higher (or lower) than the range of expected excitation frequencies. If we must increase the resonance frequency, the equation tells us our options. We can increase the stiffness of the structure, reduce its mass, or utilize a combination of the two. Correct? Moreover, it identifies how much the stiffness-to-mass ratio must be increased. The square root relationship tells, for example, that in order to increase resonance frequency by a factor of 3, we must increase the stiffness-to-mass ratio by a factor of $3^2$. This is a whale of a lot of information from a simple expression. We cannot derive nearly as much from graphs or tables.

## LINE GRAPHS

A graph of circumference vs. diameter is a straight line. We know this from the first equation. Without this knowledge, however, we could only say that it *appears* to be a straight line, suggesting a linear relationship. We can *speculate* that it holds for all values of the diameter, but this could be a dangerous assumption! Suppose I showed you a graph between $x = 0$ and $x = 50$ for the function $y = 1/(x - 5000)$. Would you see anything on the graph to suggest that $y$ will approach an infinitely large value at $x = 5000$?

Continuing with the graph of $c = \pi d$, we can read the corresponding coordinates of circumference and diameter at several points and estimate their constant of proportionality to be *about* 3.2. But this is too imprecise for most problem solving. Consider a person machining rollers or cylinders. Diameter can easily be controlled to a ten-thousandth of an inch or better. But what if circumference is the parameter we must control? Unless we know the constant of proportionality to a corresponding precision we cannot calculate the required diameter. The equation comes to our rescue because it tells us the constant is precisely $\pi$.

Graphs become much more cumbersome when the relationships are nonlinear, e.g., the variables are raised to different powers. Because of their curvature, it takes a trained eye to recognize that a circle's area curve ($\pi d^2/4$) portrays a square relationship and a sphere's volume curve ($\pi d^3/6$) a cubic relationship. However, this information is obvious in their equations.

These graphs have only two variables. But the equation for a resonance frequency has three variables. For every combination of $k$ and $m$, there is a corresponding resonance. Therefore the graph will be three-dimensional. If we plot $k$ and $m$ on the X and Y-axes, frequency will appear on the Z-axis as a surface. Can you visualize this? Look at it this way. The totality of all coordinates of $k$ and $m$ values completely fills the X-Y plane, and each coordinate has a corresponding Z-axis value, to form a surface. Welcome to the real world where we usually encounter more than three variables, which cannot be shown in a graph.

Graphs are more useful for plotting long-term trends (Chapter 7) for which a visual image is more important than obtaining precise values.

## TABLES AND BARGRAPHS

Table presentations have even more limitations than line graphs because they leave big gaps between their data points. Refer to the comparison of temperature scales in Chapter 22. You couldn't use it as a conversion table for general use, could you? But tables do an excellent job of displaying several related items. Bar graphs are a form of table that also perform well in this regard. And when the data represent breakdowns of the whole, pie charts are ideal.

THE LITTLE BROWN BOOK

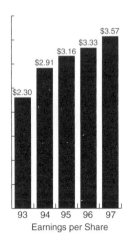
Earnings per Share

Sales Breakdown

## EMPIRICAL DATA

In the real world, equations become complicated by more than just the accumulation of variables, such as length, width, radius, force, mass, frequency, time, temperature, voltage and resistance, or by power functions, trigonometric functions, integrals and differentials, etc. Empirical data are also introduced, which add to the number of terms. They are derived from measurements or observations made with the goal of upgrading unknowns to the realm of the known or realistically predictable and thereby converting raw scientific and social data into parameters that give basic relationships a much higher degree of confidence.

Empirical data have been with us from the beginning of mathematics. We noted that the Greeks proved in the $5^{th}$ century B.C. that the area of a circle is proportional to the square of its radius. Two hundred years later they had narrowed their estimate of the ratio to between 3 10/70 and 3 10/71. Until it was defined precisely, mathematicians used these empirical values for what we now call $\pi$. Much of what we know in mathematics started this way. We shall work with empirical data in almost every profession as long as we seek new knowledge.

The field of economics is a fertile one in which to get our feet wet. Consider the matter of taxation. It sounds pretty simple. But is it? We are all familiar with sales tax. Let's say it is 6%. In a stable economy, we expect tax revenue to be directly proportional to the gross sales of taxable merchandise:

$$\text{Revenue} = 0.06 \times \text{Gross Sales}$$

Now assume your state government announces its intent to increase the sales tax rate. How, or will, this action affect revenue? As economists ponder the size of the increase, may they assume a linear relationship between revenue and tax rate? In other words, may they use this equation:

$$\Delta\text{Revenue} = \Delta\text{Tax Rate} \times \text{Gross Sales}$$

which indicates that a given increase in tax rate will produce a proportionate increase in revenue, meaning that gross sales will be unaffected by the tax increase? Or, will taxpayer behavior modify this ideal relationship?

As you ponder the question, you might first observe that many people live on a fixed income. So if sales taxes increase, they will have less money for purchases. Can you see that the above equation will not apply to them? Why? Because gross sales is a variable in their segment of the population and tax revenue will increase by a smaller percentage than the tax increase. But the equation is valid for affluent families. Or is it?

We must consider another factor. Neighboring states with lower tax rates will steal business from your state. People will cross the state line to buy a $1000 refrigerator. After a 1 percentage point increase in sales tax, your state might expect to increase revenue from $60 to $70 on each refrigerator purchase. Instead it will lose the $60 it previously collected when all residents bought at home. And while a family is in the neighboring state, they will save a bit more by purchasing gasoline, clothing and other taxable items. This was a severe problem for the province of Ontario, Canada, in recent years until it increased customs surveillance. Also, with an expanded customer base and greater profit margins, neighboring border merchants can reduce prices and attract even more business. To compound the problem, the home state will also lose income tax revenues because its merchants will make less profit and employ fewer people. The greater the tax increase, the worse the problem.

Along come economic consultants to advise the state on the amount of increase that will maximize revenue, yet prevent it from rolling over and declining. The equation for their "economic model" will contain numerous terms to account for parameters that include percent of population on fixed income, tax differentials with neighboring states, percent of residents within easy driving distance of border states and the percent of them who will bother to make cross-border purchases. The model will reflect years of accumulated data.

Similar problems confront manufacturers of consumer products and cartels that control commodity production. Prices are a function of production, sales costs, supply, demand and profit. When grocery prices are set too high, many consumers turn to alternative store brands. If the price structure of new automobiles is too high, consumers delay purchases or purchase less expensive models or fewer options. Your electrical utility company switches fuel as frequently as the relative prices of oil, gas and coal change.

Examples of cost/demand relationships go on and on. Economists say that the cost/demand curve is "inelastic" meaning there is not a straight-line relationship between selling price and sales revenue or between tax rate and tax revenue. Mathematicians say it is "nonlinear."

Equations, graphs and tables all have their rightful uses. Tables give us trends and other comparative information. Graphs give us a picture. But equations give us all the facts.

\* \* \*

## STATEMENTS OF ACCURACY

This is not the immediate subject, but it is important in the technological world. We have an ingrained habit of misusing the word accuracy. We say that a measurement or a value has an accuracy of ±1%. What does this mean to you, literally. I hope you say that it is terribly inaccurate. In that case, why boast about it? In the engineering world, if ±1% accuracy is good, ±1/2% accuracy is even better! Of course, we mean *inaccuracy* or *uncertainty*. I prefer the latter. How did this misuse ever begin? It is utter nonsense. But it is very difficult to overcome. Please help.

Now let me point out a common grammatical error. The term "data" pops up in this chapter. Many people treat the word "data" as a singular noun. But it is plural. It is a Latin word: datum is singular and data is plural. We are so used to seeing, "The data is good" that "The data are good" sounds improper. That is unfortunate. But we have made more strides in overcoming this error than we have with the misuse of the word accuracy.

One more thing, please do not buy into the popular argument that grammatical errors and misuse of words are not important because the only thing that matters is that people understand what you mean. That is ridiculous. If you don't believe it, ask a lawyer.

# 26

# ENIGMAS OF SPACE, TIME AND MATTER

We are all creatures of life's experiences, are we not? Consider a simple example. Our personal scales of magnitude are a function of our individual experiences and from time to time we change the scale. When we are very young and have spending money for the first time, we value things in terms of how many dimes or quarters they cost. As we become older and get our first odd jobs, our monetary unit increases. How many dollars or tens of dollars does a new sweater cost? When we get a full-time job, the unit advances to a hundred and then a thousand dollars. We conveniently change our monetary scale to keep it relative to our financial resources. Why, then, should we be surprised when we encounter conditions of relativity in the outside world?

## RANDOM THOUGHTS ON SPACE AND TIME

In the physical world, our scales of magnitude are much more complex than in our private world. Let's ponder distances. We long ago selected the yard or the meter as our unit of distance. It suits us well around our home. But when we travel by car or airplane, we switch to the mile or kilometer. This scale is satisfactory for travel across Earth or even out to the moon or the sun. You shouldn't have difficulty perceiving the sun as being 93 million miles away. Beyond that, it begins to get fuzzy, so we adopted the light-year as our unit for astronomical data. It may or may not improve our concept of distance, but it certainly makes the numbers easier to handle.

Earth has never experienced any serious collisions during human history. Only a few asteroids have marred our voyage. We have to conclude that space is a vast void. In *Cosmos*, Carl Sagan tells us that if you or I were randomly inserted into space, the chance would be less than 1 in $10^{33}$ that we would find ourselves near a planet! Now, what if I tell you that a void also exists at the opposite end of the scale where atoms are interacting? Physicists tell us that our bodies and all other solid matter contain 99.999% empty space. Just because we cannot see this space doesn't mean it does not exist. To view an example of this, examine a magazine photograph with your magnifying glass. What do you see? A pattern of dots and a lot of white space, correct? And you thought the images and colors were solid, didn't you?

Our basic scales for size are limited by our body characteristics. In Chapter 13 we learned that as we study smaller and smaller objects we need shorter and shorter wavelengths to examine them. Our vision's shortest wavelength enables us to see some insects, but not all of them. And we certainly cannot see germs or microbes. But when we shift to shorter wavelengths and examine things with electron microscopes, we get magnificent pictures even of crystals and atomic structures. Objects are 'small' only relative to our scale of vision.

Our most fundamental unit for measuring elapsed time is the length of time it takes Earth to rotate on its axis, because this rotation creates the day/night cycle by which we live. If we lived on Jupiter, our day would be less than ten Earth-hours; on Venus it would be several Earth-weeks.

If an airplane passes overhead at an altitude of 1000 feet and a speed of 175 MPH, we perceive it as traveling relatively fast. Increase its altitude to 10,000 feet, and it appears to be flying much slower. Turning our attention to the stars, they don't seem to change their relative position and haven't seemed to do so during recorded history. But, of course, they do. It's just that they are so far away that we cannot perceive these changes in a mere few thousand years. In fact there are constellations in which some of the so-called members are not even traveling with the others. They are just passing by. For example, only five of the seven stars in the Big Dipper are traveling together as a family.

We have the same time warp with geological events. Faults classified by geologists as active might produce a severe earthquake on average every two hundred years. But by comparison to our life spans, you and I consider the faults inactive.

Now, let's combine distance and time in yet another way. Assume you are gazing out the window of a Boeing 747 when suddenly another 747 passes a mile away heading in the opposite direction. The pilot announced after takeoff that you would be traveling at 500 MPH. You assume the other 747 is traveling at about the same speed, so the relative speed of the two aircraft was about 1000 MPH. You just experienced what it would be like to be passed at 1000 MPH at that distance if you were standing still. You experience similar events every time you travel on a highway.

But what about a situation where you are sitting in a train stopped inside a metro station. You look across at a person sitting in a train stopped on the adjacent track. Suddenly that person begins to slowly move past you. Is your train moving, or is it the other train that is moving? I hope you say you don't know. Only when you begin to feel vibrations or when the trains clear each other and you are free again to view stationary objects like billboards, can you determine which train is moving. I hope you have this experience sometime. The feeling is quite eerie.

With all this discussion of the strange things we uncover when we contemplate events on a relative basis, we should not be surprised at what Albert Einstein told us about the passage of time.

## EINSTEIN'S RELATIVITY

Until the beginning of the 20th century, scientists believed that time was absolute, i.e., every event could be labeled by a number called "time" and that all precision clocks would agree on the time interval between events. This was one of Sir Isaac Newton's tenets. Moreover, time and space were considered to be unconnected – one did not influence the other. Then, along came Einstein!

Albert Einstein rode a tram every day to his work as an examiner at the Swiss Patent Office in Berne. Above the central station was a large clock tower. He would often recall the thoughts he had when he was only sixteen years of age. "What would the world look like if I could ride on a beam of light? Suppose I were moving away from the clock tower on the very same light beam by which I am reading the clock." He reasoned that the clock would be frozen in time!

Think about it. Assume the clock read precisely 11 o'clock when he departed. If he were to travel for one second at the speed of light, it would also take one second for the clock's reading to travel that distance, because he and the clock reading would be traveling together in the same frame of reference. To him the clock would continue to read 11 o'clock no matter how far he and the light beam traveled together. In traveling at the speed of light, he would have cut himself off from the passage of time!

In a sense this seems so obvious that you wonder why no one thought of it sooner. This was a part of Einstein's genius. He could raise innocent questions that had awesome answers. This one had revolutionary implications. One is that if velocity affects the passage of time at *one* speed, it must affect it at *every* speed. All we need is the equation. And as we have seen so often, the relationship is very simple:

$$\Delta t = \Delta t_0 \sqrt{1-(v/c)^2}$$

As velocity $v$ approaches the speed of light $c$, the elapsed time $\Delta t$ approaches zero. Can you see that every speed has its own elapsed time, however small its difference from the elapsed time $\Delta t_0$ when a body is at rest? Of course, the difference – we call it *time dilation* – is significant only when a body approaches the speed of light. What is the ratio of $\Delta t$ to $\Delta t_0$ at 10% of the speed of light? My calculator says 99.5%. Not until 87% of the speed of light is elapsed time cut in half.

Einstein published these ideas in 1905 in his famous paper "The Electrodynamics of Moving Bodies." He was only 26 years of age and for it he eventually received the Nobel Prize. It became known as his Special Theory of Relativity. He even proposed an experiment to test his theory, by predicting that a clock at Earth's equator runs slower than an identical clock at the North or South Pole, because of its greater rotational speed around Earth. In those days clocks were not sufficiently accurate to measure such tiny differences. Fifty years later, scientists upheld Einstein's theory using cesium clocks placed at the center and rim of a rotating disk. It was later proved again aboard airplanes.

This means that people living in Algeria on the prime meridian have a different time than people in Greenwich, England. In the full technical sense, there is no such thing as Greenwich Mean Time! Every point on the globe has a unique time coordinate, just as it has a unique space coordinate. If you could carry a cesium clock with you and synchronize it in the morning with ones carried by your parents, each clock would read a different time at the end of the day. Unless scientists take these facts into account, they cannot compare measurements taken aboard satellites speeding through space.

In 1916 Einstein published more ideas in what we call his General Theory of Relatively. Here he included gravity which was absent from his earlier theory. One idea was that time runs slower near a massive body like Earth. Today this knowledge is important in the design of navigation systems, called Global Positioning Systems, based on signals from earth satellites.

We now accept the fact that time is not completely separate from space, but is combined with it to form what we call "space time." In other words, time is a *fourth* dimension.

## MASS-ENERGY RELATIONS

Let's return to 1905 when Einstein published his famous paper on the electrodynamics of moving bodies. Later that year, he added a postscript to it. This little gem stated that mass $M$ and energy $E$ are equivalent:

$$E = Mc^2$$

This simple formula had a tremendous implication – the possibility of converting matter into energy. Since '$c$,' the speed of light, is such a huge value, there is potential for converting a tiny amount of matter into an enormous amount of energy. Converting one gram of matter into pure energy releases enough heat to convert 34 billion grams (7.5 million gallons) of water into steam. Matter is really "frozen energy." This idea led to a quantum leap in scientific advancement that is far from being played out even today.

How could scientists implement what Einstein's equation suggested? Certainly not by common chemical reactions because the quantity of matter converted to energy (heat) is virtually immeasurable. To reap these advantages, the reactions would have to convert significant amounts of matter and do it almost instantaneously.

The wherewithal to implement Einstein's idea was not available in 1905. But after the discovery of the atomic structure (see Chapter 15), scientists began to turn to the atom itself.

When the heavy elements were formed eons ago by combining lighter elements, the mass of the new atoms was greater than the mass of the constituents from which they were created. This meant that in the process large amounts of energy had been converted into mass. If scientists could now reverse the process and cause atoms of heavy elements such as uranium or plutonium to break apart, the sum of the masses of the newly formed particles would be less than the mass of the parent particles. The missing mass would be converted to energy – rapidly and in huge quantities.

Not all scientists agreed that the atom could be split. As late as 1933, Rutherford (Chapter 15), who conceived the model of the atom, maintained that the idea was "moonshine." After hearing about this remark, Leo Szilard said, "It suddenly occurred to me that if we could find an element whose atoms could be split by neutrons and would emit two neutrons when it absorbed just one, such an element if assembled in sufficient mass could sustain a nuclear chain reaction."

With World War II approaching, news came out of Germany in 1938 that Otto Hahn had succeeded in splitting the nucleus of the uranium atom! The race toward development of an atomic bomb had begun. Early in 1941 at the University of California, Professor Glen Seaborg bombarded a millionth of a gram of plutonium with neutrons which split some of the atoms and released their energy. We call this a "fission" reaction. Once again, an Einstein theory was upheld. The next year the United States initiated the Manhattan Project and Enrico Fermi at the University of Chicago achieved the world's first controlled atomic chain reaction. American scientists won the race in 1945, and the war with Japan was terminated almost immediately. The atomic bomb is estimated to have saved more than two million lives among those of us poised for the invasion or defense of the Japanese mainland, far exceeding the casualties of the bombing. Today we use *con-*

*trolled* nuclear fission reactors in a peaceful way to produce huge quantities of electricity and thereby reduce the use of air-polluting fossil fuel. And we employ miniature reactors to power spacecraft crisscrossing our solar system.

In the creation of light elements, nature's process works in reverse of what it does with heavy elements. When two hydrogen nuclei join with neutrons to form helium, called a "fusion" reaction, the mass of helium is slightly less than the sum of the nuclear particles that joined to form it. Mass is converted to energy. This is how our sun and most other stars are powered. It is also how the hydrogen bomb operates, a mere 2 pounds of hydrogen mass being converted into the equivalent of 40 billion pounds of TNT.

Fusion is preferable as a potential source of electrical power because it does not produce radioactive wastes as does fission. The ultimate fusion fuel is deuterium, an isotope of hydrogen found in ordinary water. The amount of deuterium (1/250 ounce) present in a gallon of fresh water could produce energy equivalent to 300 gallons of gasoline. It is said that there is sufficient deuterium in the top 10 feet of Lake Michigan to supply our country's energy needs for 15,000 years! Think of it! A single power plant could supply the entire electrical needs of Canada, Mexico and the United States. It is of profound importance to the economics and to the environment of the entire world. Scientists are working feverishly on the development of controlled nuclear fusion, but to date they have been unable to economically achieve the enormous temperatures needed to set off the process.

It was no wonder, then, that the scientific world was agog when in 1989 researchers at the University of Utah announced they had observed fusion in a very simple "cold" process. No one can yet describe the kind of fusion that is occurring, but like the sun it produces helium and tritium from hydrogen and it produces more energy than it absorbs. But unlike hot fusion, it does not emit neutrons. Electrochemists are searching for an explanation. Most say there is an apparent nuclear phenomenon involved. Others say that cold fusion is impossible, but they qualify their opinion by saying it is based only upon what we know *today* about the behavior of matter and nuclei. We must be patient. After all, superconductivity was discovered in 1911, was not explained until 1972, and is just beginning to find commercial application.

Science works in two ways. Scientists like Albert Einstein use inductive logic to theorize things and then leave it to someone else to substantiate and implement and sometimes to disprove. Other scientists observe events and then employ deductive logic to search for an explanation. The latter is taking place right now with cold fusion. What is also going on is a battle between those who believe that we thoroughly understand the physics of fusion and that it can only take place at super-high temperatures and those who are willing to search for a monumental breakthrough in knowledge. It is unlikely that we have reached the ultimate in our knowledge of this or any other field.

## 27

# IS ANYONE OUT THERE?

To rephrase an old saying: Breathes there a man with soul so dead who never to himself has said, "Do UFOs really exist?" Are we the only intelligent life in the Universe? Do advanced technological civilizations exist elsewhere? If not, did they ever exist? Such questions excite every inquisitive mind. Considering the billions upon billions of stars in the Universe, wouldn't it be unrealistic to assume that we on planet Earth are unique? We are back again to the world of probabilities, aren't we?

I began my search for answers by visiting Rachel, Nevada. You reach this tiny community by driving 150 miles north from Las Vegas through the narrow valleys of Nevada's high-desert country dotted with lakes, alfalfa fields and cattle feeding stations, over highways that are perfectly straight and void of traffic as far as the eye can see. Rachel was the takeoff point to view the Groom Lake base (called "Dreamland" by military pilots) – the famous Area 51 – where top-secret military aircraft are tested.

More to the point of this story is the belief by some observers that Flying Saucers are developed or captured alien craft evaluated at a more remote facility nearby called Papoose Lake. If that were not enough, UFO sightings were reported in past years near Rachel at the famous "Black Mailbox."

A day at Rachel's Little A-Le-Inn made the trip worthwhile. I asked several of the hangers-on how to estimate the probability of intelligent life existing on other planets. "Look up the Drake equation," they promptly replied. Then a fellow who was doing research for a book on astronomy got my attention when he recited the equation from memory. He was also kind enough to tell me I could find it in *Cosmos*.

The equation was developed by Frank Drake at Cornell University to give a crude estimate of $N$, the number of advanced technological civilizations that might exist in our Milky Way galaxy. The values of the various parameters are open to discussion, of course. Let's use those (except for $f_L$) assigned by Sagan.

$$N = N_* \times f_p \times n_c \times f_l \times f_i \times f_c \times f_L$$

where:

$N_*$ = Number of stars in Milky Way. Astronomers have a good estimate of this number, $\approx 4 \times 10^{11}$

$f_p$ = Fraction of stars having planetary systems, $\approx 1/3$

$n_c$ = Number of planets suitable for life, in typical planetary system, $\approx 2$

$f_l$ = Fraction of $n_c$ on which life actually arises, $\approx 1/3$

$f_i$ = Fraction of $f_l$ on which intelligent life develops

$f_c$ = Fraction of $f_i$ on which technologically advanced civilization evolves ($f_i \times f_c \approx 1/100$)

Now, let's pause and see where we are. We have an estimate of the total number of planets in the Milky Way that might have developed during their lifetime a technological civilization equal to or more advanced than ours. But look at the values that went into our calculation! The only one that really counts is the number of stars in the Milky Way. Make adjustments to the others and you still come out with $10^{11}$, give or take a factor of ten, or somewhere between ten and a thousand trillion planets. But we have yet to deal with the final parameter:

$f_L$ = Fraction of its lifetime that a planet experiences its technologically advanced civilization

The entries above were educated estimates based upon our knowledge of astronomy. But we cannot begin to estimate $f_L$ because we do not even know its value for Earth. The problem, of course, is that we do not know whether advanced civilizations perpetuate themselves as long as their planet survives, or whether they eventually self-destruct. For forty years we feared we might destroy our own civilization by atomic war. That danger appears to have passed. Now we are concerned that we will be destroyed by damage to the ozone layer. We just don't know how long a lifetime our civilization will enjoy.

All we know is that Earth has been around for a few billion years and is expected to last a few billion more. But our civilization has fulfilled the criteria for an advanced technology for only about fifty years. If we survive another thousand years, $f_L$ will have a value of about $1000/10^{11}$. If this were typical for other civilizations, we could expect about 100 advanced technological civilizations to have developed at some time. The longer civilizations last, the greater is our estimate of their number. Then of course comes the question of how many of these might be in existence during Earth's era.

The bottom line is that there is a reasonable prospect of a few advanced civilizations existing today in our galaxy. And it is realistic to expect them to be engaged in space exploration. NASA's response was to use radio telescopes to search for intelligent transmissions. The most ambitious project, Search for Extraterrestrial Intelligence (SETI), was canceled in 1994 and continues on a limited basis through private funding.

## UFO PROSPECTS

Discovering intelligent signals is a far cry from a UFO landing in your backyard. Once again, we must consider the distances and times involved in space travel. We learned in Chapter 1 that Alpha Centauri (sun's nearest star) is about 4.2

light-years from Earth. At 6 trillion miles per light-year, that is about 25 trillion miles. And it is about 7000-times further to the center of the Milky Way.

Apollo cruised at about 5000 MPH during a major part of its journey to the moon. Let's see, that is about 1.4 miles/second (only 1/135,000 of the speed of light) so it would take an unimaginable length of time to exit our solar system.

Have we missed some things that could make the prospects for a space voyages more realistic? You bet we have. Can you identify any? The most obvious is that civilizations more advanced than ours might be able to travel faster. Ideas are floating about for developing nuclear reactors that by 2050 could be capable of accelerating a spacecraft to 10% of the speed of light. This would greatly reduce our transit time, enabling us to cover a light-year of distance every ten ship-years. We could get to Alpha Centauri in about 42 years and to Sirius in 88 years.

More exotic is the possibility of freezing our cosmonauts into a state of hibernation and then reviving them before reaching their destination. Hibernation occurs in many species in nature. And there is a laboratory in Oakland, CA that freezes corpses, with the hope of reviving them after cures are found for the diseases that killed them. Hibernation and high-speed travel could be well within the capability now of more advanced civilizations.

## CAN EINSTEIN HELP?

Can he ever! Assume that others already have, or we can develop in a few hundred years, a propulsion technology to accelerate spacecraft to speeds approaching the speed of light. Time dilation would take the place of hibernation! We would accomplish tremendous speeds and in so doing greatly diminish the aging process. This could be our best solution to the challenges of interstellar travel. Enter $v = 0.99c$ into Einstein's equation in Chapter 26, and $\Delta t/\Delta t_0$ turns out to be 0.14. Correct? This means that one Earth-year would be compressed into 0.14 ship-year. In other words, time and aging in Earth's frame of reference would be reduced by a factor of 7. Astronauts could make the trip to Sirius and back in nine years and find upon return that they had aged only 15 months in Earth's frame of reference.

This chapter has given us an emotional roller coaster ride. We started out with high hopes for the existence of UFOs. Then the Drake equation dampened our enthusiasm. Finally, the possibility of exotic new propulsion technology makes their prospects more realistic.

We have made dramatic progress since eighty-five years ago when two brothers strayed from their bicycle business and perfected powered flight. Since then we have landed humans on the moon, put space probes on many of our sister planets and sent probes into outer space. We shall continue to extend these impressive accomplishments. In the meantime we are on the lookout for civilizations that might be further along the path to interstellar travel than we are.

You might ask if anyone would be willing to take the risks of such a journey. I think this is answered in the affirmative by those who pioneered our sea routes, explored our jungles and polar regions, flew our first airplanes and ventured to the moon. Men and women are ever willing to take risks while breaking new frontiers.

*". . . The diversity of the phenomena of Nature is so great, and the treasures hidden in the heavens so rich, precisely in order that the human mind shall never be lacking in fresh nourishment."*
–Johann Kepler

\* \* \*

## DID UFOs CRASH IN NEW MEXICO?

We cannot leave this subject without some coverage of a happening in New Mexico. On July 8, 1947 the Intelligence Office at Roswell Army Air Field announced that it had possession of a 'disc' that had landed the prior week on a remote sheep ranch near Corona, New Mexico. But within three hours the story was cancelled and replaced by one that described the object as merely a radar reflector from an errant weather balloon. However, the military's behavior for weeks thereafter belied its revised story.

The military's problem was that the crash site was discovered by a ranch foreman and a young neighbor who were checking for damage after a violent thunderstorm the previous night during which they had also heard mysterious sounds. He and his family didn't think too much about what they had found because their isolated existence made them unaware of the flying saucer craze that was sweeping the nation that summer. But he was fully aware that the material he found that comprised the structure was like nothing he had ever seen. "Out of this world," might describe it. It had the appearance of tinfoil, but it could not be torn or cut or burned, was extremely lightweight and would resume its original shape after being wrinkled. Moreover, pieces were strewn over an area about 500 ft. square, an unlikely circumstance for a reflector from a helium-filled balloon which could not explode.

Even more remarkable was discovery about the same time of another crash site on the Plains of San Agustin, about 150 miles southwest of Corona. Here too, civilians were the first on the scene, in the persons of an engineer for the U.S. Soil Conservation Service and an archeology professor with a team of five students. They described dead and live humanoid bodies near a crashed saucer.

When these happenings were reported to the military, the areas were sealed off for miles around, scoured of every last piece of debris and all evidence quickly transported by truck to awaiting airplanes and delivered to military laboratories in the East and Midwest.

There was rationale at the time for the military's reaction. After all, the atom bomb had been developed and first tested in this area just a few years before. The saucers could have been a secret operation by the Soviet Union. But we now know this was not the explanation.

To this very day, the government will not release any meaningful documentation on these events. Yet we know that something very extraordinary (Crash at Corona) occurred that night in 1947. It might have been the event of the millennium!

# 28
# CHAOS AND THE UNIVERSE

Let's begin by playing a "What If?" game. Have you ever asked yourself if your life would be different if you had been born a month earlier or later? Astrologists believe it would. Or, would you be a different person if your mother were Queen of England? Would you be more comfortable? Of course. Would you be better or happier? Not necessarily. These are subtle questions and we will return to them. But first, let's apply some "What If?" questions to nature.

In this book we continually marvel at all the order we find throughout our Universe. The basic facts are irrefutable, but we are entitled to ask some "What If?" questions about the ideas generally extrapolated from them.

We learned (Chapter 6) that the discoveries of Galileo, Copernicus and Kepler established that Earth and its sister planets travel in elliptical orbits around the sun. Then Newton identified his three laws of motion and his law of gravity to explain not only planetary motion but all motions that occur in the Universe.

These men lived in Western Europe as it was emerging from the Dark Ages – a time when men's minds were influenced by the idea of a single overarching explanation for the Universe. So when the laws of motion were explained and finally accepted, everyone rushed to the conclusion that this wondrous order was preordained at the time of "Creation." This leap of faith pleased everyone, but it was bad science because the conclusion was a self-fulfilling prophesy.

What if they had sought alternate theories? Our opening questions posed two possibilities for what makes you what you are. The first addresses astrology which assumes that your personality was preordained on the day of your birth. The second assumes that you are a result of your environment and perhaps your heritage. Having considered these alternatives, isn't it fair to ask if there isn't at least one alternative explanation for the law and order that governs our Universe?

Let's go into the country and take a walk along your favorite river. The water flows smoothly and orderly. As we ponder this serene wonder of nature, is it logical to assume that we would find the same friendly waters if we followed the river all the way to its source 100 miles upstream. Well, it's possible to investigate, isn't it? So, let's hike upstream. After walking a mile or two we begin to observe some turbulence in the waters. The farther we walk, the worse it gets. Amazingly, the effects of turbulence were observed by the Greek philosopher Heracleitus while sitting along a riverbank in the ancient town of Ephesus in what is now Turkey. The waters were turbulent in that area and Heracleitus was tossing twigs into the water. As he watched them float irregularly downstream, he remarked, "Twice into the same river you could not enter." In other words, the water continually changed.

Soon we reach a waterfall where the action is really hectic. We pass the waterfall and again find placid conditions. It immediately dawns on us that the origin of the law and order we first observed was not at the creation of the river. Rather, it evolved from the turbulent waterfall!

Isn't that interesting? Law and order does not have to be preordained. It can evolve from naturally occurring, disorderly phenomena that we call CHAOS! How does this affect our conceptualization of the Universe? First of all, it is possible that the law and order we experience evolved from the chaos of the Big Bang. Secondly, if there had been more than one chaotic event, each could have evolved differently. The astounding thing about this is that there might be other universes of which we are yet unaware that developed systems of law and order vastly different from ours. Visionary astronomers give serious thought to this.

What if the likes of Galileo and Newton had been so insightful? Or what if their discoveries had been made by the ancients who attributed the various wonders of nature to individual deities? Or by the Orientals with their holistic perspective that one cannot understand a part of the Universe without first understanding the whole? (They would have been closer to being on the right track.)

Since Galileo and Newton, the search for regularity has been paramount. We disregard bits of messiness that interfere with neat pictures. We often seem to be more interested in the parts than in the whole. A common anecdote is an experiment in which a technician finds a measurement ratio of 2.001:1 one day, 2.002:1 the next day, and 1.995:1 the next, etc. A classical physicist would search for a theory to explain a perfect 2:1 ratio, while seeking ways to improve his measurement accuracy. A modern physicist would consider that there might very well be a phenomenon in the process to explain the day-to-day variation.

At any rate, we came to expect simple, predictable explanations. For instance, Galileo explained the physics of the motion of a pendulum. Well, not really, because he disregarded nonlinearities introduced by friction and air resistance when he theorized that a pendulum's period is independent of its bob's amplitude and weight. But we always have air resistance except in space and in vacuum chambers. And friction is always present with moving parts; otherwise we could produce perpetual motion machines. To make matters worse, the effects of air resistance and friction are a function of the weight of the bob and the lighter it is, the greater the effects.

This is not to denigrate Galileo's work. His was a critical first step. If not he, someone else would have accomplished it. But his ideal equations did not work perfectly in the real world.

## CHAOS STUDY

Today, scientists have mathematical tools and computational capabilities to study the messiness long bypassed by classical physicists. This new field of study, called Chaos Theory, originated in 1960 when Edward Lorenz, in attempting to predict long-term global weather patterns, discovered that the slightest changes (like one part in one thousand) in initial conditions inputted to his computer model caused enormously different predictions. This gave rise to the so-called Butterfly Effect, the notion that a butterfly stirring the air today in Chicago can affect next month's weather in Paris.

Turbulent conditions such as our waterfall also exhibit chaos, as do fluctuations in wildlife populations, changing cloud patterns, flags flapping in the wind and the behaviors of aircraft, oil flowing in pipelines and blood flowing through your heart. We apply chaos study to all types of problems. In most we try to overcome chaos. In others we utilize chaos.

Here is an interesting problem in the aircraft industry. Jet engines depend upon turbulence to mix fuel and air quickly and efficiently. If designers want to increase cruise speed, they face the problem that turbulence decreases as air speed increases. So they have to find ways to increase turbulence in the mixing process. Failure to solve this problem is one reason the supersonic Concorde is not economically viable. While this *use* of turbulence is going on, engineers have to *overcome* turbulence along the engine's interior surfaces to prevent overheating.

Mathematicians also find combinations of chaos and order in a single process. This can occur when a system exhibits both short-term and long-term variability. For example, the price of grain changes unpredictably during the course of any trading day. But it rides on a cyclical pattern that repeats seasonally, disturbed occasionally by excessive temperature or rainfall.

The study of chaos is the global study of *whole* systems. We now recognize that chaos is more to be expected than unexpected in our life experiences. We are surrounded by systems that exhibit both stable and unstable behavior. We find order obscured by chaos, as well as chaos obscured by order. Deeper understandings of chaos will benefit us all.

\* \* \*

### Thoughts on Nonlinearities

With all the mention of nonlinearities throughout this book (such as what we said about Galileo's work in this chapter), it is well to clarify what the term means. First, we define a system as being "linear" if its elements are related in a proportional manner. Scales and thermometers are linear devices because their readings are proportional to weight and temperature. But most systems are "nonlinear" because their elements are not related in this manner.

Suppose your father is a salesman working for a salary plus commission or depends upon a year-end bonus for a portion of his income. This situation is nonlinear because his sales and the company's profits are a function of such things as the national and global economies, product competitiveness (which in turn is dependent upon the company's productivity, engineering skills and overall management capability) and competitors' performances. Your father's sales ability plays a minor role in all of this, yet these events heavily influence his income. Also see discussion of empirical data in Chapter 25. Such hodgepodges of continually changing relationships are what nonlinearities and chaos are all about.

## 29
# DEMON dB

We have a demon in our mathematics closet! It might be better to leave it there. But we won't, because it is a very useful tool. Its name is "decibel" – abbreviated dB. You surely recognize the name from its use with stereo equipment or acoustics. It had its origin with Alexander Graham Bell, the inventor of the telephone, who is honored by the "bel" in decibel. We must begin our story, however, with the bel itself.

Much of Bell's laboratory work was devoted to measuring ratios between electrical power in two telephone lines or between two signals on the same line. These ratios ranged from tiny fractions up to millions. Finding it unwieldy to use such wide-ranging values directly, he sought a more manageable way to report them. He needed to translate them into a much smaller set of values – a code, so to speak.

His first step was to adopt the logarithmic scale, which gave him equal spacing for each 10-fold increase or decrease in ratio. Then he conceived the idea of creating a bel scale to label each increment of 10-times. If one measurement of electrical power was 10× another, he said it was one bel greater; 100×, 2 bels; 1000×, 3 bels, etc. Similarly, −1 bel for 1/10, −2 bels for 1/100, etc. You can see from the scale that the number of bels is identical to the exponent of 10 in the ratio it

| Power: | $10^{-2}$ | $10^{-1}$ | $10^0$ | $10^1$ | $10^2$ | $10^3$ | $10^4$ | $10^5$ |
|---|---|---|---|---|---|---|---|---|
| Ratio: | 1/100 | 1/10 | 1 or Ref. | 10 | 100 | 1000 | 10,000 | 100,000 |
| bel: | −2 | −1 | 0 | +1 | +2 | +3 | +4 | +5 |
| dB (re Ref.) | −20 | −10 | 0 | +10 | +20 | +30 | +40 | +50 |

THE LITTLE BROWN BOOK

*Alexander Graham Bell (1847-1922), opening the NewYork-Chicago telephone line.*

represents. If you examine the graduations on these two scales, you will see that Alexander Graham Bell had succeeded in translating a large range of values on a logarithmic scale to a small set of linear values on the bel scale.

The bel was fine and dandy except that it was much too large an increment for most work. It was like a grocer trying to sell strawberries by the bushel, instead of the quart. So, being good disciples of the decimal system, scientists soon divided the bel into ten divisions, called decibels (dB).

By now, you are probably asking for the code, i.e., the formula for converting ratios into decimal notation. Before doing this, however, let's develop our story one step further. Bell soon recognized that if he substituted a reference (ref) value for ×1 on the ratio scale, he could extend his application of the dB to include the absolute measurement of electrical power. If his reference were one watt (W), for example, then 20 dB re 1W would represent 100 W. If his reference were 1 mW, then 100 W (100,000 mW) would be 50 dB re 1 mW.

Now we are ready to reveal Bell's code. Once again, the underlying mathematics are easy:

$$\text{Power (bel)} = \log \frac{W_1}{W_2} \quad \text{or} \quad \log \frac{W_1}{W_{\text{ref}}}$$

$$\text{Power (dB)} = 10\log \frac{W_1}{W_2} \quad \text{or} \quad 10\log \frac{W_1}{W_{\text{ref}}}$$

The following abbreviated table is readily expandable to tenths of a dB. It enables us to convert from decibels to

| Pressure Ratio | Power Ratio | − dB + Ratio | Power Ratio | Pressure Ratio |
|---|---|---|---|---|
| 1.0000 | 1.0000 | 0 | 1.000 | 1.000 |
| 0.8913 | 0.7943 | 1.0 | 1.259 | 1.122 |
| 0.7943 | 0.6310 | 2.0 | 1.585 | 1.259 |
| 0.7079 | 0.5012 | 3.0 | 1.995 | 1.413 |
| 0.6310 | 0.3981 | 4.0 | 2.512 | 1.585 |
| 0.5623 | 0.3162 | 5.0 | 3.162 | 1.778 |
| 0.5012 | 0.2512 | 6.0 | 3.981 | 1.995 |
| 0.4467 | 0.1995 | 7.0 | 5.012 | 2.239 |
| 0.3981 | 0.1585 | 8.0 | 6.310 | 2.512 |
| 0.3548 | 0.1259 | 9.0 | 7.943 | 2.818 |
| 0.3162 | 0.1000 | 10.0 | 10.000 | 3.162 |
| 0.1778 | 0.03162 | 15.0 | 1.620 | 5.623 |
| 0.1000 | 0.01000 | 20.0 | 100.000 | 10.000 |
| 0.03162 | 0.00100 | 30.0 | 1000.000 | 31.623 |

absolute ratios or vice versa. When the denominator is a reference value, the dB participates in defining an absolute value, but the dB itself still represents a ratio. The following example of an amplifier's signal and noise outputs ties the two uses together. Be sure to reference the power columns on the right hand side of the above table re 1 mW.

Signal Level $(3.16\text{ W}) = 3.16 \times 1000$ mW
$= (5 + 30)$ dB re 1 mW
$= 35$ dB re 1 mW

Self - Noise Level $(1.26\text{ mW}) = 1$ dB re 1 mW

Signal - to - Noise Ratio $= (35 - 1)$ dB $= 34$ dB
(from above equations)

or $= \dfrac{3.16 \times 1000 \text{ mW}}{1.26 \text{ mW}}$
$= 2.51 \times 1000 = 34$ dB
(from table)

## NON-POWER MEASUREMENTS

We have been careful so far to limit our discussions to measurements of power, specifically power generated by an amplifier in an electrical circuit. We could also have included acoustical and other forms of mechanical power, but we still would not have covered all of Alexander Graham Bell's needs for the decibel. For example, when evaluating the performance of his telephone's microphone, he measured its output voltage. This was not a power measurement. Yet he still had use for the advantages of the decibel notation. What to do? Well, how do electrical power and voltage relate?

$$\text{Watt (power)} = E(\text{volt}) \times I(\text{current})$$
$$I = E/R (\text{resistance})$$
$$\therefore W = E^2/R$$

Here he had a kind of apples and oranges problem. Power is proportional to voltage-squared, not just voltage. How could he adapt voltage measurements to the decibel notation? Since he had ordained that the dB always be reserved for power ratios, he had to use voltage-squared in order to conform to his ground rules. Hence:

$$\text{Voltage (dB)} = 10\log\left[\frac{V_1}{V_2}\right]^2 \text{ or } 10\log\left[\frac{V_1}{V_{\text{ref}}}\right]^2$$

But the minute mathematicians saw this, they brought the exponent outside the ratio and rewrote the formula as:

$$\text{Voltage (dB)} = 20\log\frac{V_1}{V_2} \text{ or } 20\log\frac{V_1}{V_{\text{ref}}}$$

Wow! Now 20 dB represented a 10-fold change, instead of 100-fold. We commonly call it a "pressure" dB. A new conversion table was in play. It is included with the power dB in the preceding table.

We must live with the fact that 'dB' has one set of values for power ratios, and a different set for voltage ratios. Similarly, this occurs in acoustics, between power and pressure ratios. Most acoustic data are in pressure decibels. A prominent acoustics consultant once wrote, "Unfortunately, long ago the acousticians borrowed a unit (the dB) from the electrical engineers, and we have lived to regret it." Do you begin to understand why I questioned the wisdom of letting the dB out of our closet? It is a terribly confusing notation for anyone not schooled in the basics presented here.

It gets even worse. Try convincing a person living next to a freeway that a noise reduction of 6 dB will reduce sound pressure by 50%. He wants that "92 dB sound level cut in half to 46 dB" – a level you might experience in a private office. Worse yet, the human ear marches to its own drummer. It perceives a 10 dB change as being half (or twice) as loud. Notwithstanding these idiosyncrasies, the decibel notation is a very valuable tool.

## OTHER USES OF THE DECIBEL

Besides being a convenient code for translating a wide range of values into a small set of numbers, we also use it in the hi fi industry as a measure of deviation when specifying frequency response. Are you into hi fi? If you examine the specifications for your stereo amplifier or speaker, you may find it rated 100 Hz to 20 kHz (±1 dB), meaning that its output is constant within ±1 dB over the range of 100 Hz to 20 kHz. Or, it might be rated as 50 Hz to 22 kHz (–3 dB), meaning its output is attenuated by 3 dB at 50 Hz and 22 kHz, the "cut-off" frequencies beyond which output rolls off very rapidly. In each case, you should know whether the manufacturer is specifying power or pressure dB.

We also use the decibel as a surrogate for percentage. We

might specify an accuracy of ±1 dB – almost always the equivalent of a pressure ratio. So it stands for about ±12%, doesn't it? And ±0.1 dB stands for about ±1.2%. These are close enough to 10% and 1% for most mental calibrations. Whenever you see the decibel used in this manner, chalk up another victory for the use of percentages.

## EARTHQUAKES

Did I say we never used the bel? Not quite, but I came close. My apologies to Professor Charles F. Richter who in 1935 at the California Institute of Technology developed the Richter scale because he was tired of newsmen asking him to compare the magnitudes of earthquakes.

Professor Richter observed that the energy (power) released by earthquakes spans a range of at least $10^{10}$ to 1, an ideal place to apply the bel notation. But he must also have realized that the general public would never recognize the meaning of the bel, so he did not mention it when he labeled his scale from 1 to 10. The Richter numbers for earthquakes are reported to the nearest tenth of a unit, e.g., "5.7 on the Richter scale." That would be 5.7 bels re his reference. Only occasionally does a news announcement include the statement that each increment of 1 on the Richter scale represents a 10-fold increase or decrease in magnitude.

We should be quick to observe that the Richter number does not tell us how severe an earthquake was in our locality. It merely estimates the energy released at the epicenter. Since earthquakes spread out in all directions and distribute their energy over greater and greater areas (as do sound waves), the farther we are from the epicenter, the less the severity. You would never imagine this, judging from news broadcasts, would you? So much for show biz.

## 30
# FREQUENCY ANALYSIS

We all enjoy music. And you may even play a musical instrument. The simplest example of such an instrument is a tuning fork – a mechanical device constructed so that when excited mechanically, e.g., by tapping it against your knee, it vibrates at one frequency. I've had one in my desk for years. Its frequency is 1024 Hz. That's two octaves above middle C. If I strike it and then hold it near my ear, I hear its tone as it sets air molecules into motion at 1024 Hz. Similarly, if I hold it against my desk, I can feel the resulting vibration.

Trumpets, pianos, clarinets, trombones and other instruments operate in a similar fashion, except that you can change the tone and its amplitude and you can produce several tones simultaneously. We have a name for this process. It is called synthesis. Does it occur to you that if you had a full set of tuning forks covering the ranges of all the instruments, and if you could turn them on and off at will and control their amplitude, you could create music without using any of the traditional musical instruments? Indeed, the famous 19$^{th}$-century German acoustician, Hermann von Helmholtz, demonstrated this very thing using electrically driven tuning forks. We have equivalent devices today – called electronic music synthesizers – used to produce many of the background sounds you hear on television. Electronic pianos and organs are also in this class.

### FREQUENCY ANALYSIS

When listening to music, you immediately identify the

songs, even the artists. And when you hear a discordant note, it really gets your attention. Why is this so? It is because your brain analyzes the frequencies and amplitude patterns present in the sounds. This is the exact opposite of synthesis.

We encounter frequency analysis in our daily lives. If our car is not running smoothly, the mechanic listens carefully to the engine and its accessories and gently feels the vibrations. Doctors use a stethoscope to listen to our heartbeat and they feel our pulse. Both individuals analyze sounds and vibrations with which they are intimately familiar, but they would find it impossible to write an instruction that would enable you or me to perform the same analysis.

In industry we must continuously monitor the vibration of critical machinery. And hospitals must continuously monitor patients in intensive care. Periodic human examinations are inadequate for these requirements. We need to automate the analysis process. Guess what? The mathematical model for frequency analysis was given to us around 1800, but we waited 180 years for development of appropriate instrumentation to implement it on a mass scale.

We all know that every function, such as a musical tone or the vibration of a machine, can be portrayed as a graph. But what about the reverse situation? Can every graph be defined by a function? Does every graph have an equation? Although all mathematicians in those days knew that some graphs could be represented by a combination of sine waves, they thought them to be special cases. Only the French mathematician, Jean-Baptiste Fourier (1768-1830), recognized that every periodic curve (one which repeats itself regularly) can be represented by a series of sine waves, each with its characteristic amplitude and each frequency being a multiple of the fundamental frequency. We call it a Fourier series or expansion:

$$y = A_0 + A_1 \sin \omega t + A_2 \sin 2\omega t + \ldots A_n \sin n\omega t$$

The values $A_n$ are called Fourier Coefficients. The series is written in the form where omega is the function's fundamental frequency ($\omega = 2\pi f$ radians per second). All component frequencies are multiples of the fundamental frequency. The objective is to identify prominent frequencies and their amplitudes and to use this information to help us evaluate the signatures of machinery and human bodies. We said the function had to be periodic. If it is not, Fourier had a solution for that too – include a corresponding cosine expansion.

I do not expect it to be obvious that all of a machine's component vibrations are multiples of its rotational (fundamental) frequency. It surely wasn't obvious to Fourier's fellow mathematicians. However, it becomes more clear when you consider that all the operations going on in a machine are mechanically or structurally linked to the main shaft. So, up and down the frequency range every troublesome vibration is a multiple of the rotational frequency. Other components are minor. Enough of this detail. Trust me, it works.

Fourier gave us the now-familiar formula. But computing the coefficients (amplitudes of the component frequencies) was a horrendous task, so his work had no everyday use. That never bothers pure scientists, does it? Eventually its time would come. And it did around the 1940s when vacuum tube technology gave us our first modern instrumentation.

First came the radio. It is a form of frequency analyzer, you know, because it enables us to tune through the range of electromagnetic radiation frequencies in which radio signals are broadcast. Switch to vibration frequencies and you have a vibration analyzer which can tune in each vibration component and measure its amplitude. This was a gigantic step because it enabled us to implement Fourier's work on a daily basis. But it was no utopia, because analog analysis is very slow and it is difficult if not impossible to separate closely spaced frequencies. You know that from tuning your radio dial.

Twenty years later, computers came on the scene and gave us digital analyzers which overcame the deficiencies of analog analyzers. Computers sample the instantaneous amplitudes of waveforms, prior to the number-crunching analysis process. And this presented a problem. As sampling rates became faster and faster to accommodate higher and higher frequencies, the number of computer operations was expanding as the square of the number of samples. For each 10-times increase in sampling rate (frequency), computer operations increased by a factor of 100-times. This tremendously limited the application of computers to Fourier analysis. It might have been acceptable for vibration, which is relatively low in frequency, but it was intolerable for noise and a host of other high-frequency applications.

To the rescue in 1965 came J.W. Cooley and J.W. Tukey of Bell Laboratories, who devised a new algorithm which, for $N$ samples, reduced the number of operations from $N^2$ to $2N$. They succeeded in substituting linear growth for exponential growth! It was reported at the time that for $N = 8192$ data points, computing time on a mainframe computer was reduced from 30 minutes to 5 seconds. The rest is history.

Today we have hand held analyzers whose microprocessors analyze noise and vibration in real-time. And we are on the brink of voice-activated computers (most assuredly using some form of Fourier analysis) that will greatly improve accuracy and reduce communication time between humans and computers. Another giant step in human development.

# 31
# CALCULUS 001

No, calculus was not created for the purpose of grinding you down. Its purpose is to help you solve practical problems. Calculus was invented independently in the late 17th century by Sir Isaac Newton and Gottfried Wilhelm von Leibnitz. It consists of two parts – integral calculus which deals with quantities that change over time, and differential calculus which deals with instantaneous rates of change of quantities. Instruction invariably begins with differential calculus, but I find it much easier to develop the logic by beginning with integral calculus.

## INTEGRATION

We need go no further than our automobile to get an introduction to calculus. Let's assume we are embarking on a trip and will want to know how far we have driven at any time. But our car's odometer is broken! How can we estimate the total distance driven? I'll show you. Fortunately the speedometer is working. If we could drive at constant velocity, the problem would be trivial, wouldn't it? Common sense would tell us to simply multiply our velocity (mi/hr) by the number of hours driven. That's pretty obvious. But it leads to a very important observation! We have in reality calculated the area under a curve of velocity vs. time (in this case a long rectangle) and discovered that the area is a measure of distance.

Our problem is more complex, however, because we will be continuously changing speed. Our record of velocity might look something like the following illustration. What scheme would you suggest to estimate the distance traveled at any

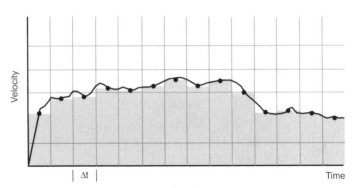

time? I hope you say, "Divide the time axis into a series of short intervals, and record the speedometer reading for each interval. Then calculate the area of each small rectangle, and sum all the areas." Congratulations. You have taken the first step in developing the concept of integral calculus.

Let's translate your instructions into a formula, which is usually the best way to give instructions. The incremental rectangles have a variable height $V_N$ and a fixed base of $\Delta t$. Therefore:

$$\text{Distance} = \sum_{0}^{T} \left( V_1 \Delta t + V_2 \Delta t + \ldots V_N \Delta t \right) = \sum_{0}^{T} \left( V_1 + V_2 + \ldots V_N \right) \Delta t$$

Your estimate will naturally have some error because your speedometer readings were not very representative of average velocity in intervals during which velocity changed rapidly. Practicality limited your intervals to perhaps five minutes. To improve accuracy, a modern car's trip computer might sample once each second. In calculus we make the interval infinitesimally short, i.e., the rectangles approach line segments which intersect the curve at an infinite number of points. We have a notation for this – instead of $\Delta t$, we write $dt$ (pronounced "dee-tee"), or more generally $dx$ (dee-ex) for any X-axis unit of measurement. These are called "differentials." We also replace the summation symbol with a stylized 'S' integration symbol.

Finally we enter the curve's actual equation if we know it.

$$\text{Distance} = \int_{T_1}^{T_2} f(t)\,dt$$

Otherwise we symbolize it by $f(t)$ or $f(x)$. And, of course, we can integrate between any two times. We do not have to start with $t = 0$ or $x = 0$. This completes the fundamental concept of integral calculus.

## DIFFERENTIATION

Now, let's change our problem. Assume we want to measure velocity at any instant, but the speedometer is not working. We must find a way to use the odometer. All right, we know that velocity is the distance traveled over a given period of time, so we might record the distance (Y-axis) traveled in the last ten seconds. We can estimate this distance by observing the increase in odometer reading over this period of time. Let's bring calculus into play. We make the time interval infinitesimally small. What happens is that the distance also becomes infinitesimally small. Now we divide one by the other. This gives us $dy/dt$ or $dy/dx$ – our symbol for differentiation. It represents the slope of the curve. Wonderful! The derivative of a curve at any point is a measure of the rate of change of the variable. In this example, it tells us the rate of change of distance, or the velocity.

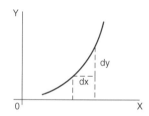

Have you noticed that when we integrate velocity we get distance, and when we differentiate distance we get velocity?

This tells us that integration and differentiation are inverse operations – either operation cancels the other. We can extend our scenario to include acceleration (rate of change of velocity). Do you see that acceleration is the first derivative of velocity and the second derivative of distance? And that you can integrate acceleration once to get velocity or twice to get distance? It does not matter which one we actually measure, calculus enables us to calculate the other two.

## A BRIEF HISTORY

Time to take a break and reflect on some history. Newton was the first person to discover calculus. It has been called his most important contribution to mathematics. "Fluxions" – meaning flowing or fluctuating – was his name for it. Strangely, he kept his powerful new tool a secret, although he used it privately to develop other findings. Pretty sneaky. He did not publish it until Leibnitz made his own announcement years later. Newton never believed that Leibnitz had conceived of calculus independently. A quarrel developed among their followers and unfortunately spread to the two great men themselves. Leibnitz eventually overshadowed Newton in the calculus arena because he had a far superior system of notation. It became the standard we use today. Newton's notation is obsolete.

## PRACTICAL APPLICATIONS FOR EQUATIONS

The problems we contrived in order to introduce integration and differentiation involve what we call "empirical" data, i.e., data that we obtain from measurements, such as a physical record of velocity vs. time. We are not interested in knowing the equation of the curve. All we want to do is use the curve itself to calculate areas and slopes. Got that?

A second type of situation is one in which we have a formula. Every mathematical function, be it a power, trigonometric function, logarithm, etc., has derivatives and integrals. Calculus textbooks are loaded with them, complete with instructions on how to derive them.

Many are quite simple. If I told you that $d(x^2)/dx = 2x$ and that $d(x^3)/dx = 3x^2$, you would recognize the pattern immediately and tell me – I hope – that $d(x^n)dx = nx^{n-1}$. I will also tell you that the derivative of a constant is zero. Let's apply these bits of information. I'll bet you know that the equation for a straight line is $y = mx + b$. To graph the line, you could substitute two values for $x$, solve for the corresponding values of $y$, and then draw a straight line between the two points. Calculus gives you an easier and more direct way. The slope is the derivative of $mx + b$, correct? So the slope is $m$. And we know that $b$ is the Y-axis intercept. You can solve this problem by simply inspecting the equation and then drawing a straight line with slope $m$ through point $b$ on the Y-axis.

If I told you that $d(\sin x)/dx = \cos x$, you would know that the derivative of a sine wave is a cosine wave (another sine wave but with different phase). Let me ask you what is the maximum value of a sine wave. If you don't know, you can consult a reference table and look up the highest value you can find for a sine wave. With calculus, the question is easier to answer. Picture a sine wave in your mind. Observe that it will have its highest value when its slope is a horizontal line – in other words, when its slope, or derivative, is zero. Great! Now you know not only that the derivative of $\sin x$ is $\cos x$, but that $\sin x$ is maximum when $\cos x = 0$. In your reference table you can go directly to the angle whose cosine is zero, and the sine of that angle is the maximum value of a sine wave. You can also work this out on a calculator. Are you familiar with "arc cos," i.e., "the angle whose cosine is . . ." function? My calculator says that 90° is the angle whose cosine is zero (i.e., arc cos 0 = 90°). Therefore, the maximum value of a sine wave is at 90°, and that value is 1. This may seem involved.

But, believe me, calculus is a savior when dealing with complex equations. Equating the derivative to zero is the way we find maxima and minima of any function.

One more function. The derivative of $e^x$ is $e^x$. It is the only function known that is its own derivative. In other words, its rate of growth is identical to its current value. This helps explain some of the things we said in Chapter 16.

Let's replay a couple of things. Repeated differentiation eventually reduces a power function to zero. With sin $x$ and cos $x$, we get alternating cosine and sine functions. And $e^x$ does not change at all. If you like variety, what more could you want? If you like Calculus 001, you'll love Calculus 101.

In science and engineering we have endless problems similar to the ones described. For example, an electrical generator is less efficient at low and high loads than at medium loads – just as your car gets better mileage at medium speeds. At what load is the generator most efficient? If we have an equation expressing generator efficiency as a function of electrical load, we can take the derivative, set it equal to zero and solve for the output load. In most cases, the hard part is obtaining the equation. The calculus is quite easy.

Now, some examples of integration. What is the total force that a lake exerts on a dam? The deeper the lake, the greater the force exerted by the water above. Start with an equation of force vs. depth. Write the integral of the equation; substitute in it the maximum and then the minimum depth; and subtract the resulting values. Another example. If we return to the Gaussian distribution, we can write its equation, integrate it and calculate the area (percentage) between any two values of sigma.

Are you beginning to see that calculus simplifies otherwise difficult or even impossible problems of evaluating changing quantities? This is what makes calculus so powerful a tool for scientists, engineers and economists.

# CONVERSION TABLES

### International System of Units (SI)

**Basic Units**

| Quantity | Unit | Symbol |
|---|---|---|
| Length | metre | m |
| Mass | kilogram | kg |
| Time | second | s |
| Electric Current | ampere | A |
| Thermodynamic Temperature | kelvin | K |
| Amount of Substance | mole | mol |
| Luminous Intensity | candela | cd |

## English System of Units

The English system of units is inherently confusing. In everyday American life, the *pound* (abbreviated lb) is used either as a force or as a weight, both having the units of force. To further complicate matters, some technical writers seek to provide a system parallel to the Metric system by using either of two quantities for force and either of two corresponding quantities for mass. Thus, some adopt the pound as the unit of force and define a *slug* as the unit of mass. Others define a *poundal* as the unit of force and adopt the pound as the unit of mass.

The English system is still used in U.S. commerce and industry and both the Metric and English systems are used in U.S. engineering education. Charlie Jackson, a well-known turbomachinery consultant, once said, "Americans are going metric 'inch-by-inch'." Charlie also gave us some additional bits of wisdom, "The three most basic equations are: $E = IR$, $F = Ma$, and Don't PUSH on a Rope!"

**Consistent Systems of Units.** Two consistent systems of units are commonly used, the *mks* and the *fss* systems. To describe them, let us start with Newton's second law.

$$\text{force} = \text{mass} \times \text{acceleration} \qquad (1)$$

In the *meter-kilogram-second* system

$$\text{no. of Newtons} = \text{no. of kg} \times \text{no. of m/s}^2 \qquad (2)$$

In the *foot-slug-second* system

$$\text{no. of pounds force (lb)} = \text{no. of slugs} \times \text{no. of ft/s}^2 \qquad (3)$$

---

The Conversion Factors section is based on appendices from *Noise and Vibration Control*, edited by Leo L. Beranek. Copyright © 1971 by McGraw-Hill, Inc., New York, NY. Revised Edition © 1988 by Leo L. Beranek. Published by the Institute of Noise Control Engineering, Washington, D.C.

The relationships between the magnitudes of the units are:
$$1 \text{ kg} = 2.205 \text{ lb weight}$$
$$1 \text{ kg} = 0.0685 \text{ slug}$$
$$1 \text{ slug} = 14.59 \text{ kg}$$
$$1 \text{ Newton} = 0.225 \text{ lb force}$$
$$1 \text{ lb force} = 4.448 \text{ Newtons}$$
$$1 \text{ slug} = 32.17 \text{ lb weight}$$
$$1 \text{ lb weight} = 0.03108 \text{ slug} = 0.454 \text{ kg}$$

Example 1: If 1 kg is to be accelerated 1 m/s$^2$, we see, by Eq. (2), that a force of 1 Newton is required. How many pounds (force) is required for the same result?

Solution: 1 kg equals (2.205/32.17) slug and 1 m/s$^2$ = 3.28 ft/s$^2$. Thus, by Eq. (3), 0.225 lb (force) is required.

**Inconsistent Systems of English Units.** Two inconsistent systems of English units are commonly encountered, the *fps* and the *ips* systems.

In the *foot-pound-second* system

no. of pounds force (lb) = [no. of pounds weight (lb)/g] × no. of ft/s$^2$    (4)

where g is the acceleration due to gravity in units of ft/s$^2$, that is, 32.17 ft/s$^2$.

In the *inch-pound-second* system

no. of pounds force (lb) = [no. of pounds weight (lb)/g] × no. of in./s$^2$    (5)

where g is the acceleration due to gravity in units of in./s$^2$, that is, 386 in./s$^2$

Mechanical engineers often use the *in.-lb-sec* system in the field of shock and vibration.

Example 2: One kilogram is accelerated five meters per second$^2$. Find the force necessary to do this in Newtons and pounds (force).

Solution:

1 kg = 2.2 lb weight = 0.0685 slug

5 m/s$^2$ = 16.4 ft/s$^2$

F (Newtons) = 1 × 5 = 5 Newtons

F (lb) = 0.0685 × 16.4 = 1.124 lb (force)

**Conversion Factors.** The following values for the fundamental constants were used in the preparation of the conversion factors:
$$1 \text{ m} = 39.37 \text{ in.} = 3.281 \text{ ft}$$
$$1 \text{ lb (weight)} = 0.4536 \text{ kg} = 0.03108 \text{ slug}$$
$$1 \text{ slug} = 14.594 \text{ kg}$$
$$1 \text{ lb (force)} = 4.448 \text{ Newtons}$$
$$\text{acceleration due to gravity} = 9.807 \text{ m/s}^2 = 32.174 \text{ ft/s}^2$$
$$\text{density of } H_2O \text{ at } 4° C = 10^3 \text{ kg/m}^3$$
$$\text{density of Hg at } 0° C = 1.3595 \times 10^4 \text{ kg/m}^3$$
$$1 \text{ U.S. lb} = 1 \text{ British lb}$$
$$1 \text{ U.S. gallon} = 0.83267 \text{ British gallon}$$

# Conversion Tables

| To Convert | Into | Multiply by | Conversely, Multiply by |
|---|---|---|---|
| acres | $ft^2$ | $4.356 \times 10^4$ | $2.296 \times 10^{-5}$ |
| | $mile^2$ (statute) | $1.562 \times 10^{-3}$ | 640 |
| | $m^2$ | 4,047 | $2.471 \times 10^{-4}$ |
| | hectare ($10^4 m^2$) | 0.4047 | 2.471 |
| atm | in. $H_2O$ at 4° C | 406.80 | $2.458 \times 10^{-3}$ |
| | in. Hg at 0° C | 29.92 | $3.342 \times 10^{-2}$ |
| | ft $H_2O$ at 4° C | 33.90 | $2.950 \times 10^{-2}$ |
| | mm Hg at 0° C | 760 | $1.316 \times 10^{-3}$ |
| | $lb/in.^2$ | 14.70 | $6.805 \times 10^{-2}$ |
| | $Newtons/m^2$ | $1.0132 \times 10^5$ | $9.872 \times 10^{-6}$ |
| | $kg/m^2$ | $1.033 \times 10^4$ | $9.681 \times 10^{-5}$ |
| °C | °F | (°C × 9/5) + 32 | (°F − 32) × 5/9 |
| cm | in. | 0.3937 | 2.540 |
| | ft | $3.281 \times 10^{-2}$ | 30.48 |
| | m | $10^{-2}$ | $10^2$ |
| circular mils | $in.^2$ | $7.85 \times 10^{-7}$ | $1.274 \times 10^6$ |
| | $cm^2$ | $5.067 \times 10^{-6}$ | $1.974 \times 10^5$ |
| $cm^2$ | $in.^2$ | 0.1550 | 6.452 |
| | $ft^2$ | $1.0764 \times 10^{-3}$ | 929 |
| | $m^2$ | $10^{-4}$ | $10^4$ |
| $cm^3$ | $in.^3$ | 0.06102 | 16.387 |
| | $ft^3$ | $3.531 \times 10^{-5}$ | $2.832 \times 10^4$ |
| | $m^3$ | $10^{-6}$ | $10^6$ |
| deg (angle) | radians | $1.745 \times 10^{-2}$ | 57.30 |
| dynes | lb (force) | $2.248 \times 10^{-6}$ | $4.448 \times 10^5$ |
| | Newtons | $10^{-5}$ | $10^5$ |
| $dynes/cm^2$ | $lb/ft^2$ (force) | $2.090 \times 10^{-3}$ | 478.5 |
| | $Newtons/m^2$ | $10^{-1}$ | 10 |
| ergs | ft-lb (force) | $7.376 \times 10^{-8}$ | $1.356 \times 10^7$ |
| | joules | $10^{-7}$ | $10^7$ |
| $ergs/cm^3$ | $ft-lb/ft^3$ | $2.089 \times 10^{-3}$ | 478.7 |
| ergs/s | watts | $10^{-7}$ | $10^7$ |
| | ft-lb/s | $7.376 \times 10^{-8}$ | $1.356 \times 10^7$ |
| $ergs/s-cm^2$ | $ft-lb/s-ft^2$ | $6.847 \times 10^{-5}$ | $1.4605 \times 10^4$ |
| fathoms | ft | 6 | 0.16667 |
| ft | in. | 12 | 0.08333 |
| | cm | 30.48 | $3.281 \times 10^{-2}$ |
| | m | 0.3048 | 3.281 |

| To Convert | Into | Multiply by | Conversely, Multiply by |
|---|---|---|---|
| $ft^2$ | $in.^2$ | 144 | $6.945 \times 10^{-3}$ |
| | $cm^2$ | $9.290 \times 10^2$ | 0.010764 |
| | $m^2$ | $9.290 \times 10^{-2}$ | 10.764 |
| $ft^3$ | $in.^3$ | 1728 | $5.787 \times 10^{-4}$ |
| | $cm^3$ | $2.832 \times 10^4$ | $3.531 \times 10^{-5}$ |
| | $m^3$ | $2.832 \times 10^{-2}$ | 35.31 |
| | liters | 28.32 | $3.531 \times 10^{-2}$ |
| ft $H_2O$ at 4° C | in. Hg at 0° C | 0.8826 | 1.133 |
| | $lb/in.^2$ | 0.4335 | 2.307 |
| | $lb/ft^2$ | 62.43 | $1.602 \times 10^{-2}$ |
| | Newtons/$m^2$ | 2989 | $3.345 \times 10^{-4}$ |
| gal (liquid U.S.) | gal (liq. Brit. Imp.) | 0.8327 | 1.2010 |
| | liters | 3.785 | 0.2642 |
| | $m^3$ | $3.785 \times 10^{-3}$ | 264.2 |
| gm | oz (weight) | $3.527 \times 10^{-2}$ | 28.35 |
| | lb (weight) | $2.205 \times 10^{-3}$ | 453.6 |
| hp (550 ft-lb/s) | ft-lb/min | $3.3 \times 10^4$ | $3.030 \times 10^{-5}$ |
| | watts | 745.7 | $1.341 \times 10^{-3}$ |
| | kW | 0.7457 | 1.341 |
| in. | ft | 0.0833 | 12 |
| | cm | 2.540 | 0.3937 |
| | m | 0.0254 | 39.37 |
| $in.^2$ | $ft^2$ | 0.006945 | 144 |
| | $cm^2$ | 6.452 | 0.1550 |
| | $m^2$ | $6.452 \times 10^{-4}$ | 1550 |
| $in.^3$ | $ft^3$ | $5.787 \times 10^{-4}$ | $1.728 \times 10^3$ |
| | $cm^3$ | 16.387 | $6.102 \times 10^{-2}$ |
| | $m^3$ | $1.639 \times 10^{-5}$ | $6.102 \times 10^4$ |
| kg | lb (weight) | 2.2046 | 0.4536 |
| | slug | 0.06852 | 14.594 |
| | gm | $10^3$ | $10^{-3}$ |
| $kg/m^2$ | $lb/in.^2$ (weight) | 0.001422 | 703.0 |
| | $lb/ft^2$ (weight) | 0.2048 | 4.882 |
| | $gm/cm^2$ | $10^{-4}$ | 10 |
| $kg/m^3$ | $lb/in.^3$ (weight) | $3.613 \times 10^{-5}$ | $2.768 \times 10^4$ |
| | $lb/ft^3$ (weight) | $6.243 \times 10^{-2}$ | 16.02 |
| liters | $in.^3$ | 61.03 | $1.639 \times 10^{-2}$ |
| | $ft^3$ | 0.03532 | 28.32 |

| To Convert | Into | Multiply by | Conversely, Multiply by |
|---|---|---|---|
| liters *(cont.)* | pints (liquid U.S.) | 2.1134 | 0.47318 |
| | quarts (liquid U.S.) | 1.0567 | 0.94636 |
| | gal (liquid U.S.) | 0.2642 | 3.785 |
| | $cm^3$ | 1000 | 0.001 |
| | $m^3$ | 0.001 | 1000 |
| $\log_e n$, or $\ln n$ | $\log_{10} n$ | 0.4343 | 2.303 |
| m | in. | 39.371 | 0.02540 |
| | ft | 3.2808 | 0.30481 |
| | yd | 1.0936 | 0.9144 |
| | cm | $10^2$ | $10^{-2}$ |
| $m^2$ | $in.^2$ | 1550 | $6.452 \times 10^{-4}$ |
| | $ft^2$ | 10.764 | $9.290 \times 10^{-2}$ |
| | $yd^2$ | 1.196 | 0.8362 |
| | $cm^2$ | $10^4$ | $10^{-4}$ |
| $m^3$ | $in.^3$ | $6.102 \times 10^4$ | $1.639 \times 10^{-5}$ |
| | $ft^3$ | 35.31 | $2.832 \times 10^{-2}$ |
| $m^3$ | $yd^3$ | 1.3080 | 0.7646 |
| | $cm^3$ | $10^6$ | $10^{-6}$ |
| microbars (dynes/$cm^2$) | lb/$in.^2$ | $1.4513 \times 10^{-5}$ | $6.890 \times 10^4$ |
| | lb/$ft^2$ | $2.090 \times 10^{-3}$ | 478.5 |
| | Newtons/$m^2$ | $10^{-1}$ | 10 |
| miles (nautical) | ft | 6080 | $1.645 \times 10^{-4}$ |
| | km | 1.852 | 0.5400 |
| miles (statute) | ft | 5280 | $1.894 \times 10^{-4}$ |
| | km | 1.6093 | 0.6214 |
| $miles^2$ (statute) | $ft^2$ | $2.788 \times 10^7$ | $3.587 \times 10^{-8}$ |
| | $km^2$ | 2.590 | 0.3861 |
| | acres | 640 | $1.5625 \times 10^{-3}$ |
| mph | ft/min | 88 | $1.136 \times 10^{-2}$ |
| | km/min | $2.682 \times 10^{-2}$ | 37.28 |
| | km/hr | 1.6093 | 0.6214 |
| nepers | dB | 8.686 | 0.1151 |
| Newtons | lb (force) | 0.2248 | 4.448 |
| | dynes | $10^5$ | $10^{-5}$ |
| Newtons/$m^2$ | lb/$in.^2$ (force) | $1.4513 \times 10^{-2}$ | $6.890 \times 10^3$ |
| | lb/$ft^2$ (force) | $2.090 \times 10^{-2}$ | 47.85 |
| | dynes/$cm^2$ | 10 | $10^{-1}$ |
| lb (force) | Newtons | 4.448 | 0.2248 |

| To Convert | Into | Multiply by | Conversely, Multiply by |
|---|---|---|---|
| lb (weight) | slugs | 0.03108 | 32.17 |
|  | kg | 0.4536 | 2.2046 |
| lb $H_2O$ (distilled) | $ft^3$ | $1.602 \times 10^{-2}$ | 62.43 |
|  | gal (liquid U.S.) | 0.1198 | 8.346 |
| $lb/in.^2$ (weight) | $lb/ft^2$ (weight) | 144 | $6.945 \times 10^{-3}$ |
|  | $kg/m^2$ | 703 | $1.422 \times 10^{-3}$ |
| $lb/in.^2$ (force) | $lb/ft^2$ (force) | 144 | $6.945 \times 10^{-3}$ |
|  | $N/m^2$ | 6894 | $1.4506 \times 10^{-4}$ |
| $lb/ft^2$ (weight) | $lb/in.^2$ (weight) | $6.945 \times 10^{-3}$ | 144 |
|  | $gm/cm^2$ | 0.4882 | 2.0482 |
|  | $kg/m^2$ | 4.882 | 0.2048 |
| $lb/ft^2$ (force) | $lb/in.^2$ (force) | $6.945 \times 10^{-3}$ | 144 |
|  | $N/m^2$ | 47.85 | $2.090 \times 10^{-2}$ |
| $lb/ft^3$ (weight) | $lb/in.^3$ (weight) | $5.787 \times 10^{-4}$ | 1728 |
|  | $kg/m^3$ | 16.02 | $6.243 \times 10^{-2}$ |
| poundals | lb (force) | $3.108 \times 10^{-2}$ | 32.17 |
|  | dynes | $1.383 \times 10^4$ | $7.233 \times 10^{-5}$ |
|  | Newtons | 0.1382 | 7.232 |
| slugs | lb (weight) | 32.17 | $3.108 \times 10^{-2}$ |
|  | kg | 14.594 | 0.06852 |
| $slugs/ft^2$ | $kg/m^2$ | 157.2 | $6.361 \times 10^{-3}$ |
| tons, short (2,000 lb) | tonnes (1,000 kg) | 0.9075 | 1.102 |
| watts | ergs/s | $10^7$ | $10^{-7}$ |
|  | hp (550 ft-lb/s) | $1.341 \times 10^{-3}$ | 745.7 |

# SUGGESTIONS FOR FURTHER READING

Barrow, J., *Theories of Everything*, Ballantine, 1991.

Boslough, J., "Searching for the Secrets of Gravity," *National Geographic*, May 1989.

Boslough, J., "The Enigma of Time," *National Geographic*, March 1990.

Briggs, J. and Peat, E., *Looking Glass Universe*, Simon & Schuster, 1984.

Bronowski, J., *The Ascent of Man*, Little, Brown, 1973.

Clark, R., *Einstein, the Life and Times*, World Publishing, 1971.

Collins, M., "Mission to Mars," *National Geographic*, November 1989.

Ferris, T. (Editor), *The World Treasury of Physics, Astronomy and Mathematics*, Little, Brown, 1993.

Ford, K., *Elementary Particles*, Blaisdell, 1963.

Friedman, S. and Berliner, D., *Crash at Corona*, Random House, 1992.

Gleich, J., *Chaos: Making a New Science*, Penguin, 1987.

Gore, R., *The Once and Future Universe*, *National Geographic*, June 1983.

Hawking, S., *A Brief History of Time*, Bantam, 1990.

Hurt III, H., *For All Mankind*, Atlantic Monthly Press, 1988.

Kippenhahn, R., *Bound to the Sun*, W. H. Freeman, 1988.

McLeish, J., *Number*, Ballantine, 1991.

Mook, D. and Vargish, T., *Inside Relativity*, Princeton University Press, 1987.

Mott, L. and Weaver, J., *Conquering Mathematics*, Plenum, 1993.

Sagan, C. and Shklowskii, I., *Intelligent Life in the Universe*, Holden-Day, 1966.

Sagan, C., *Cosmos*, Random House, 1980.

Schroeder, G., *Genesis and the Big Bang*, Bantam, 1990.

Spielberg, N. and Anderson, B., *Seven Ideas That Shook the Universe*, Wiley, 1987.

Thorne, K., *Black Holes & Time Warps*, W. W. Norton, 1994.

# INDEX

accuracy, 147
Arabic numerals, 7-8, 42
Apollo mission, 55-61
Archimedes, 22, 51
Aristotle, 26, 76
art, 93-6
atomic number, 80-3
averages, 45-9, 99-100, 113-4

bel, 169, 173-4
Bell, A.G., 169-73
Big Bang, 2-3, 20, 43
binary numbers, 9, 108-9
Bohr, N., 66-7

calculus, 179-182
carbon dating, 90-1
Celsius, A., 122
chaos, 164-8
chance, 15-16
complex numbers, 39-40, 92
conic sections, 125-34
Copernicus, N., 26-30
cycloid, 132-3

decibel (dB), 169-74
decimal system, 7-8
Descartes, R., 39
dilation (time), 152
Drake equation, 158-9
Dürer, A., 94

$e$, 39, 87-92, 183-4
earthquakes, 174
Einstein, A., 62, 74-5, 86, 151-6, 160-1
electromagnetics, 60-9
electron, 80-2, 85
ellipse, 27-8, 125-7, 131-2

empirical data, 144-6
equal-percentage scale (see logarithmic scale)
Eratosthenes, 26
escape velocity, 55, 61
ether, 62, 70-5
Euler, L., 41, 91-2
exponential growth, 87-92
exponents law, 43

Fibonacci sequence, 42
fission, 154-5
Fourier, J., 176-7
frequency analysis, 176-8
fusion, 155-6

Galilei, G., 23, 26-30, 132, 139, 166
Gauss, K., 40-7, 97
Gaussian distribution, 97-102, 105-6, 115-6
geocentric theory, 26
geometric mean, 49-50
golden ratio (section), 94-6
gram, 120
graphs, 142-4
gravity, 29-30, 56-7, 85-6
**Greek alphabet, 44**
Greenwich Mean Time, 135-6, 139-40, 154

Haldane, J.B.S., 21-2
heliocentric theory, 26-7
hexadecimal numbers, 109-11
Hippocrates, 51
Hubble space telescope, 4-5
hyperbola, 130-2

infinity, 18-20

interest, 31, 87-9
**International (SI) System of Units, 185**
isotopes, 82-3

Kelvin, Lord W.T., 122-3
Kepler, J., 26-8, 161

latitude, 135-7
law of exponents, 43
Leibnitz, G., 110, 133, 179-81
Leonardo of Pisa, 42
light, color, 64-6
   speed, 3, 71,-2,151-3
   year, 3-5
logarithms, 34, 91-2
logarithmic scale, 34-6
longitude, 135-40
lottery, 16-7

Mars mission, 30
Maxwell, J., 71
mean (see average)
median, 46-8
Mendeleev, D., 76-80
meter, 75, 118-9
Michelson, A., 71-5
Milky Way, 2-3, 158, 160
mode, 46, 48

Napier, J., 9, 34
nautical mile, 136-7
neutron, 81-3, 84-5
Newton, Sir Isaac, 30, 133, 151, 179, 181-2
nonlinearities, 145-6, 168
number
   bases, 8-10, 108-11
   types, 37-42
   odds, 15-7

organic growth (see exponential growth)

parabola, 126, 129-30,131
percentage, 31-6
percentile, 47
Periodic Table, 77-81
perspective drawing, 93-4
pi ($\pi$), 39, 51-4
polls, 103-7
probability, 11-5
Ptolemy, C., 26
Pythagoras, 38-9, 133-4

quark, 84-6

radian, 53-4
radix, 108-11
relativity, 151-3
**ratio prefixes, 44**
Richter, C. (see earthquakes)
RMS, 114-5
Roman numerals, 6-8

Sagan, Carl, 149, 158
scientific notation, 43-4
sound, 68-9
standard deviation, 100-2, 105-6
standard error, 101-2

temperature scales, 120-3

UFO, 157-63

wavelength, 63-4

yard, 117-9

*Reference tables are shown in boldface.*

## ABOUT THE AUTHOR

Anthony J. Schneider received his M.S. in mathematics from John Carroll University. After a career in the marketing of electronic measurement instrumentation, he is directing his writing skills toward inspiring students to enrich their lives through a better understanding of science and mathematics. He resides in San Clemente, CA.

## ABOUT THE PUBLISHER

Jack K. Mowry received his B.S. in engineering administration from Case Institute of Technology. He currently publishes the magazine *Sound and Vibration* which is circulated world wide to individuals involved with noise and vibration control. He resides in Bay Village, OH and is active with the Case Alumni Association.